Advances in Modelling and Simulation of Materials in Applied Sciences

Advances in Modelling and Simulation of Materials in Applied Sciences

Guest Editor

Evaggelos Kaselouris

MDPI

Basel • Beijing • Wuhan • Barcelona • Belgrade • Novi Sad • Cluj • Manchester

Guest Editor
Evaggelos Kaselouris
Music Technology and
Acoustics
Hellenic Mediterranean
University
Rethymnon
Greece

Editorial Office
MDPI AG
Grosspeteranlage 5
4052 Basel, Switzerland

This is a reprint of the Special Issue, published open access by the journal *Materials* (ISSN 1996-1944), freely accessible at: https://www.mdpi.com/journal/materials/special_issues/B7S7E84BJI.

For citation purposes, cite each article independently as indicated on the article page online and as indicated below:

Lastname, A.A.; Lastname, B.B. Article Title. *Journal Name* **Year**, *Volume Number*, Page Range.

ISBN 978-3-7258-5741-8 (Hbk)
ISBN 978-3-7258-5742-5 (PDF)
https://doi.org/10.3390/books978-3-7258-5742-5

Contents

About the Editor

Evaggelos Kaselouris

Evaggelos Kaselouris graduated from the National Technical University of Athens (NTUA) with a diploma from the Faculty of Applied Mathematics and Physical Sciences, followed by postgraduate studies in Biomedical Engineering at NTUA. He subsequently obtained his PhD from the Technical University of Crete. He is currently an Assistant Professor at the Hellenic Mediterranean University (HMU), and a member of the Research Staff of the Institute of Plasma Physics and Lasers (IPPL) of the HMU Research and Innovation Centre, with research expertise in multiphysics computational simulations. To date, Evaggelos has participated in 15 international and national research projects and has contributed to more than 145 scientific publications, 47 of which have been published in international peer-reviewed journals.

materials

MDPI

Editorial

Advances in Modelling and Simulation of Materials in Applied Sciences

Evaggelos Kaselouris [1,2]

1 Physical Acoustics and Optoacoustics Laboratory, Department of Music Technology and Acoustics, Hellenic Mediterranean University, 74133 Rethymnon, Greece; vagfem@hmu.gr

2 Institute of Plasma Physics and Lasers-IPPL, University Research and Innovation Centre, Hellenic Mediterranean University, 74150 Rethymnon, Greece

Advances in materials science, engineering, and computer science have created new opportunities for physicists and engineers to develop novel methods for material processing and characterization. This has led to a greater need for advanced modeling and simulation techniques that can capture multiphysics phenomena across various spatial and temporal scales. As a result, modeling and simulation have become essential tools, working alongside experimental measurements to deepen our understanding of materials.

In recent years, significant progress has been achieved in computational approaches such as finite element simulations [1–10], multiphysics simulations [11–18], ab initio and molecular dynamics simulations [19–25], and structural optimization algorithms [26–32]. Applications span a wide spectrum of engineering fields, including mechanical and manufacturing engineering [33–46], mechanical behavior and characterization of materials [47–60], thermomechanics [16–18,61–66] and thermohydrodynamics [67–69], vibration and acoustics [70–77], and fluid dynamics [78–84]. Materials under study range from metals and alloys to composites, metamaterials, and multifunctional structures. This enlarged bibliography highlights the breadth of modelling and simulation approaches and highlights the effective integration of computational and experimental methods, showing that simulations are not just for predicting outcomes but also for guiding the design, processing, and optimization of materials.

Within this broad context, the contributions collected in this Special Issue (Refs. [85–94]) can be grouped into five specific topics that illustrate how these general advances are applied to cutting-edge problems in materials science. These are as follows:

(i) Finite element simulations in mechanical and manufacturing engineering: Advanced roll forming of dual-phase steels [85] and optimization of corrugated tubes for crashworthiness [86].

(ii) Mechanical behavior and characterization of materials: Integration of digital image correlation for accurate property determination [87]; fibrous material remodeling under stress [88]; lightweight and multifunctional structures, including TPMS sandwich cores [89] and tunable phononic crystals [90].

(iii) Reliability and vibration performance of engineered systems: Fatigue life prediction of composite hydrogen storage vessels under random vibration [91].

(iv) Thermohydrodynamic multiphysics modelling: Simulation of plasma arc melting for titanium alloys [92].

(v) Ab initio, atomistic modelling of materials: Short-range ordering in gallium melts [93] and first principle calculations in Cu–Zn alloys [94].

Song and co-authors [85] performed finite element simulations using Abaqus to investigate the roll forming process of a wheel rim made from high-strength dual-phase steel

DP590 following flash butt welding. The study modeled the flaring, three-roll forming, and expansion stages, incorporating weld- and heat-affected zone microstructural data to analyze stress–strain distribution, thickness reduction, and forming stability. Results showed stress and strain concentrations in welded and coarse- or fine-grained heat-affected zones, with significant thickness reduction in groove and flange sections due to softening in intercritical and subcritical HAZ regions. Microstructural features, such as layered martensite and coarse grains, played a decisive role in deformation behavior. Process optimization, including a recommended friction coefficient of 0.4 and an optimal feed-to-roll speed ratio of 15 mm/s to 1:1.2, improved forming quality. The findings highlight the need for careful application of DP590 in rim manufacturing and suggest broader potential in automotive, aerospace, and rail sectors, as well as opportunities for advanced simulation methods, parameter refinement, and microstructure control to enhance performance and durability.

To address the limitations of longitudinal corrugated tubes (LCTs) and circumferential corrugated tubes (CCTs) in crashworthiness, a novel bi-directional corrugated tube (BCT) was developed in the study of Zou and co-authors [86], combining the high specific energy absorption of CCTs with the stable platform force of LCTs. Formed by rolling a bi-directional corrugated structure into a circular tubular form, the BCT was evaluated through experiments and finite element simulations, with results showing strong agreement. Parametric studies and multi-objective optimization revealed that tailored geometric parameters significantly improve energy absorption and stability, with the BCT achieving up to 80% higher specific energy absorption than CCTs and 45% higher than LCTs, while reducing ultimate load capacity. These findings position the BCT as a promising energy absorber for applications in automotive crash boxes, aerospace structures, and other fields requiring enhanced crashworthiness, with future work focusing on high-speed impact and vibration behavior.

Accurate mechanical property parameters are essential for reliable finite element simulations, and the study of Wang and co-authors [87] compared strain gauge and digital image correlation (DIC) techniques in tensile testing of 316 L stainless steel used in pressurized water reactor pipelines. Results showed that while both methods yield nearly identical data within the strain gauge's range, DIC offered a far broader measurement scope, capturing complete strain evolution and enhancing elastoplastic simulation accuracy. Finite element predictions calibrated with DIC data closely matched experimental strain distributions, validating DIC as a powerful tool for both parameter acquisition and simulation verification. Its wider applicability positions it as a promising method for characterizing complex, heterogeneous nuclear structures such as welded joints or defect-containing components, where traditional gauges face spatial limitations.

Rodella [88] presented a continuum model for fibrous materials, like those in biological tissues and composites, that undergo plastic remodeling under mechanical stress. The research identified three distinct classes of material behavior based on how their internal fibers reorient under tension. This reorientation led to complex and extreme responses, including phase segregation (where different regions of the material develop unique fiber alignments) and transitions between auxetic and non-auxetic properties. The findings, derived from a combination of analytical and numerical methods, offer significant implications for engineering new materials with tunable properties for applications ranging from soft robotics and protective equipment to biomedical implants.

Triply periodic minimal surface (TPMS) sandwich structures offer promising lightweight, high-strength, and energy-absorbing solutions, and the study of Vasile and co-authors [89] evaluated eight novel TPMS cores and a stochastic topology using finite element analysis, using Ansys, with both implicit and conventional modeling approaches. While implicit modeling proved as accurate as traditional CAD-based methods, it required up to four times greater computational effort. Sensitivity analyses identified optimal meshing parameters, and four material models for SLA-manufactured photopolymer parts were assessed, with the Multilinear model achieving the highest accuracy at only 2.54% deviation from experimental results. Numerical simulations closely matched experimental failure mechanisms for most cases, validating the predictive capability of the approach. The findings highlight implicit modeling as a viable tool for optimizing complex geometries before fabrication, offering valuable guidance for designing high-performance sandwich structures in engineering applications.

The study of Kaniolakis Kaloudis and co-authors [90] investigated how unit cell multiplicity influenced the acoustic response of phononic crystals (PCs), using a laser–plasma sound source characterization method combined with computational multiphysics simulations. Results revealed that increasing cell multiplicity deepened phononic bandgaps, exponentially attenuated acoustic energy in the first bandgap, and enhanced spectral flatness in passband regions, while also increasing the number of resonant peaks tied to spherical harmonics and standing wave modes. A methodology was proposed to determine the number of unit cells required to achieve a specific bandgap depth, especially with respect to the behavior of a structure with infinite length. These findings underscore the tunability of PCs for targeted sound control and noise insulation applications, and set the stage for future studies on defected cells, higher-order structures, and material density effects.

Xiong and co-authors [91] presented an experimental–numerical investigation of composite hydrogen storage vessels (CHSVs) under random vibration, a key reliability concern for fuel cell vehicle applications. Finite element models, validated against modal testing with <9% frequency deviation, were used to predict stress distribution and fatigue life via Steinberg's and Dirlik's methods. Results identified the head section as the critical fatigue zone, with initial maximum stresses of 469.4 MPa (winding layer) and 173.0 MPa (liner), corresponding to lifetimes of ~1.66×10^6 and 3.06×10^6 cycles. Optimization of ply parameters reduced stresses by 66% and 85%, respectively, and extended both components' lifetimes to over 1×10^7 cycles, meeting high-cycle fatigue standards. The findings provide design and optimization guidelines applicable to a broad class of CFRP–metal or CFRP–polymer pressure vessels.

The study by Bellot and co-authors [92] combined advanced numerical modeling and pilot-scale experiments to evaluate inclusion removal in titanium alloys processed via Plasma Arc Melting–Cold Hearth Remelting (PAMCHR), a key route for producing aeronautical-grade materials. Using a custom PAM3D model on the Ansys-Fluent platform, the work simulated thermohydrodynamic behavior, melting rates, and high-density inclusion (HDI) trajectories in a 100–150 kg/h pilot furnace. Multiphysics simulations, validated by crucible bath profile and residence time measurements, revealed that over 50% of torch-supplied heat is lost through crucible cooling and that bath depth reached approximately one-third of the metal height in the crucible. Both modeling and HDI insemination tests confirmed efficient sedimentation-based removal of refractory inclusions—an advantage over processes like Vacuum Arc Remelting. These results provide a predictive tool for process optimization and highlight PAMCHR's potential for reliably producing ultra-clean titanium alloys for critical aerospace applications.

Wang and co-authors [93] employed the Wulff cluster model (WCM) to elucidate the short-range ordering (SRO) structure of gallium melts, revealing clusters with Wulff-shaped exteriors and crystallographic symmetry interiors. High-temperature XRD measurements at 523 K, 623 K, and 723 K showed excellent agreement with Ab initio simulations in peak positions, widths, and relative intensities, even well above the melting point. Minor deviations at 523 K are linked to pre-nucleation onset. By integrating experimental data,

DFT-derived surface energies, and structural modelling, the study confirms WCM's capability to describe SRO in melts with partial covalent character, such as gallium, providing insight into melt structure–property relationships.

The study of Huang and co-authors [94] combined first-principles calculations with the Boltzmann transport equation to investigate electrical and thermal conductivities of Cu–Zn alloys across α, $\alpha + \beta'$, and β' phases with 0–50 at.% Zn. Results revealed a non-monotonic decrease–increase trend in both properties, consistent with experiments. Electronic structure analysis showed that Cu d-orbitals dominated near the Fermi level in the α-phase, with increasing Zn lowering the effective density of states and reducing electron transport. Thermal conductivity trends, validated via the Wiedemann–Franz law, aligned with measured data. These findings provide theoretical insight into transport mechanisms and optimization strategies for complex alloy systems.

Acknowledgments: The Guest Editor would like to thank all the contributing authors, the reviewers, and the editorial team of *Materials*.

Conflicts of Interest: The author declares no conflicts of interest.

References

1. Lee, J.-W.; Lee, M.-G.; Barlat, F. Finite Element Modeling Using Homogeneous Anisotropic Hardening and Application to Spring-Back Prediction. *Int. J. Plast.* **2012**, *29*, 13–41. [CrossRef]
2. Johnsen, J.; Holmen, J.K.; Myhr, O.R.; Hopperstad, O.S.; Børvik, T. A Nano-Scale Material Model Applied in Finite Element Analysis of Aluminium Plates under Impact Loading. *Comput. Mater. Sci.* **2013**, *79*, 724–735. [CrossRef]
3. Kaushik, V.; Narasimhan, R.; Mishra, R.K. Finite Element Simulations of Notch Tip Fields in Magnesium Single Crystals. *Acta Mater.* **2014**, *189*, 195–216. [CrossRef]
4. Mazière, M.; Forest, S. Strain Gradient Plasticity Modeling and Finite Element Simulation of Lüders Band Formation and Propagation. *Contin. Mech. Thermodyn.* **2015**, *27*, 83–104. [CrossRef]
5. Spanos, K.N.; Georgantzinos, S.K.; Anifantis, N.K. Mechanical Properties of Graphene Nanocomposites: A Multiscale Finite Element Prediction. *Compos. Struct.* **2015**, *132*, 536–544. [CrossRef]
6. Chatzopoulou, G.; Karamanos, S.A.; Varelis, G.E. Finite Element Analysis of UOE Manufacturing Process and Its Effect on Mechanical Behavior of Offshore Pipes. *Int. J. Solids Struct.* **2016**, *83*, 13–27. [CrossRef]
7. Brilakis, E.V.; Kaselouris, E.; Markatos, K.; Mastrokalos, D.; Provatidis, C.; Efstathopoulos, N.; Chronopoulos, E. Mitchell's Osteotomy Augmented with Bio-Absorbable Pins for the Treatment of Hallux Valgus: A Comparative Finite Element Study. *J. Musculoskelet. Neuronal Interact.* **2019**, *19*, 234–244.
8. Berardo, A.; Borasso, M.; Gallus, E.; Pugno, N.M. Finite Element Simulations and Statistical Analysis for the Tribological Design of Mutually Textured Surfaces and Related Experimental Validation. *J. Tribol.* **2025**, *147*, 031701. [CrossRef]
9. Langlet, A.; Petit, A.; Chaari, F.; Delille, R.; Drazetic, P. On the Response of Re-Entrant Auxetic Structures under Dynamic Crushing: Analytical Approach and Finite Element Simulations. *Int. J. Crashworthin.* **2025**, *30*, 387–401. [CrossRef]
10. Nayak, A.S.; Prieto, A.; Fernández-Comesaña, D. Accuracy of Post-Processing Projections for Displacement Based Finite Element Simulations in Room Acoustics. *Finite Elem. Anal. Des.* **2025**, *248*, 104349. [CrossRef]
11. Ionescu, V. Finite Element Method Modelling of a High Temperature PEM Fuel Cell. *Rom. J. Phys.* **2014**, *59*, 285–294.
12. Massimino, G.; Colombo, A.; D'Alessandro, L.; Procopio, F.; Ardito, R.; Ferrera, M.; Corigliano, A. Multiphysics Modelling and Experimental Validation of an Air-Coupled Array of PMUTs with Residual Stresses. *J. Micromech. Microeng.* **2018**, *28*, 054005. [CrossRef]
13. Kaselouris, E.; Dimitriou, V.; Fitilis, I.; Skoulakis, A.; Koundourakis, G.; Clark, E.L.; Chatzakis, J.; Bakarezos, M.; Nikolos, I.K.; Papadogiannis, N.A.; et al. Preliminary Investigation on the Use of Low Current Pulsed Power Z-Pinch Plasma Devices for the Study of Early Stage Plasma Instabilities. *Plasma Phys. Control. Fusion* **2018**, *60*, 014031. [CrossRef]
14. Ahmed, K.; Paul, B.K.; Jabin, M.A.; Biswas, B. FEM Analysis of Birefringence, Dispersion and Nonlinearity of Graphene Coated Photonic Crystal Fiber. *Ceram. Int.* **2019**, *45*, 15343–15347. [CrossRef]
15. Zhang, J.; Zhao, L.; Rosenkranz, A.; Song, C.; Yan, Y.; Sun, T. Nanosecond Pulsed Laser Ablation of Silicon—Finite Element Simulation and Experimental Validation. *J. Micromech. Microeng.* **2019**, *29*, 075009. [CrossRef]
16. Orphanos, Y.; Kosma, K.; Kaselouris, E.; Vainos, N.; Dimitriou, V.; Bakarezos, M.; Tatarakis, M.; Papadogiannis, N.A. Integrated Nanosecond Laser Full-Field Imaging for Femtosecond Laser-Generated Surface Acoustic Waves in Metal Film–Glass Substrate Multilayer Materials. *Appl. Sci.* **2019**, *9*, 269. [CrossRef]

17. Kaselouris, E.; Fitilis, I.; Skoulakis, A.; Orphanos, Y.; Koundourakis, G.; Clark, E.L.; Chatzakis, J.; Bakarezos, M.; Papadogiannis, N.A.; Dimitriou, V.; et al. The Importance of the Laser Pulse–Ablator Interaction Dynamics Prior to the Ablation Plasma Phase in Inertial Confinement Fusion Studies. *Philos. Trans. R. Soc. A* **2020**, *378*, 20200030. [CrossRef]
18. Tserpes, K.; Kormpos, P. Detailed Finite Element Models for the Simulation of the Laser Shock Wave Response of 3D Woven Composites. *J. Compos. Sci.* **2024**, *8*, 83. [CrossRef]
19. Scherer, C.; Schmid, F.; Letz, M.; Horbach, J. Structure and Dynamics of B${2}O{3}$ Melts and Glasses: From *ab initio* to Classical Molecular Dynamics Simulations. *Comput. Mater. Sci.* **2019**, *159*, 73–85. [CrossRef]
20. Chu, W.; Saidi, W.A.; Prezhdo, O.V. Long-Lived Hot Electron in a Metallic Particle for Plasmonics and Catalysis: *Ab initio* Nonadiabatic Molecular Dynamics with Machine Learning. *ACS Nano* **2020**, *14*, 10305–10313. [CrossRef]
21. Liu, J.; He, X. *Ab initio* Molecular Dynamics Simulation of Liquid Water with Fragment-Based Quantum Mechanical Approach under Periodic Boundary Conditions. *Chin. J. Chem. Phys.* **2021**, *34*, 761–768. [CrossRef]
22. Dinsdale, A.; Khvan, A.; Smirnova, E.A.; Ponomareva, A.V.; Abrikosov, I.A. Modelling the thermodynamic data for hcp Zn and Cu–Zn alloys—An *ab initio* and calphad approach. *Calphad* **2021**, *72*, 102253. [CrossRef]
23. Singh, M.; Shukla, A.; Chakraborty, B. An *Ab-initio* Study of the Y Decorated 2D Holey Graphyne for Hydrogen Storage Application. *Nanotechnology* **2022**, *33*, 405406. [CrossRef]
24. Dornheim, T.; Schwalbe, S.; Moldabekov, Z.A.; Vorberger, J.; Tolias, P. *Ab initio* Path Integral Monte Carlo Simulations of the Uniform Electron Gas on Large Length Scales. *J. Phys. Chem. Lett.* **2024**, *15*, 1406–1412. [CrossRef]
25. Huang, N.Q.; Shi, Y.; Meng, Z.K.; Ban, Z.G. A unified photo-excited GaAs model from *ab initio* simulation in terahertz regime. *J. Phys. D Appl. Phys.* **2024**, *57*, 265102. [CrossRef]
26. Li, H.-S.; Cao, Z.-J. Matlab Codes of Subset Simulation for Reliability Analysis and Structural Optimization. *Struct. Multidiscip. Optim.* **2016**, *54*, 391–410. [CrossRef]
27. Rong, Z.; Yi, J.; Li, F.; Liu, Y.; Eckert, J. Thermal Stress Analysis and Structural Optimization of Ladle Nozzle Based on Finite Element Simulation. *Mater. Res. Express* **2022**, *9*, 045601. [CrossRef]
28. Chen, J.; Liang, X.; Xu, P.; Yao, S. Crashworthiness Analysis and Multi-Objective Optimisation of Multi-Cell Windowed Structures under Dynamic Impact Loading. *Shock Vib.* **2022**, *2022*, 2263308. [CrossRef]
29. Xie, C.; Wang, D.; Zong, L.; Kong, D. Crashworthiness analysis and multi-objective optimization of spatial lattice structure under dynamic compression. *Int. J. Impact Eng.* **2023**, *180*, 104713. [CrossRef]
30. Li, H.; Zhuang, S.; Zhao, B.; Yu, Y.; Liu, Y. Visualization of the Gas Permeation in Core–Shell MOF/Polyimide Mixed Matrix Membranes and Structural Optimization Based on Finite Element Equivalent Simulation. *Sep. Purif. Technol.* **2023**, *305*, 122504. [CrossRef]
31. Rekatsinas, C.S.; Dimitriou, D.K.; Chrysochoidis, N.A. A Road Map to the Structural Optimization of a Type-C Composite UAV. *Aerospace* **2024**, *11*, 211. [CrossRef]
32. Jin, H.; Jin, L. Stress Analysis and Structural Optimisation of Insulating Rods Under Dynamic and Static Loads. *High Voltage* **2025**, *10*, 116–125. [CrossRef]
33. Kaselouris, E.; Papadoulis, T.; Variantza, E.; Baroutsos, A.; Dimitriou, V. A Study of Explicit Numerical Simulations in Orthogonal Metal Cutting. *Solid State Phenom.* **2017**, *261*, 339–346. [CrossRef]
34. Stender, M.E.; Beghini, L.L.; Sugar, J.D.; Veilleux, M.G.; Subia, S.R.; Smith, T.R.; San Marchi, C.W.; Brown, A.A.; Dagel, D.J. A Thermal–Mechanical Finite Element Workflow for Directed Energy Deposition Additive Manufacturing Process Modeling. *Addit. Manuf.* **2018**, *21*, 556–566. [CrossRef]
35. James, S.; De La Luz, L. Finite Element Analysis and Simulation Study of CFRP/Ti Stacks Using Ultrasonic Additive Manufacturing. *Int. J. Adv. Manuf. Technol.* **2019**, *104*, 4421–4431. [CrossRef]
36. Ibrahim, M.M.; El Aghoury, I.M.; Ibrahim, S.A.-B. Experimental and Numerical Investigation of Ultimate Shear Strength of Unstiffened Slender Web-Tapered Steel Members. *Thin-Walled Struct.* **2020**, *148*, 106601. [CrossRef]
37. Soliman, H.A.; Shash, A.Y.; El Hossainy, T.M.; Abd-Rabou, M. Investigation of Process Parameters in Orthogonal Cutting Using Finite Element Approaches. *Heliyon* **2020**, *6*, e05498. [CrossRef]
38. Saddighi, A.; Mahbod, M.; Asgari, M. Bi-tubular corrugated composite conical–cylindrical tube for energy absorption in axial and oblique loading: Analysis and optimization. *J. Compos. Mater.* **2020**, *54*, 2399–2432. [CrossRef]
39. Wang, Y.; Xu, X.; Ren, B.; Liu, J.; Zhao, R. Effect of Geometrical Factors on Torsion in Cold Roll Forming of the Lower Side Beam of a Car. *Appl. Sci.* **2021**, *11*, 7852. [CrossRef]
40. Albak, E.I. Crashworthiness design and optimization of nested structures with a circumferentially corrugated circular outer wall and inner ribs. *Thin-Walled Struct.* **2021**, *167*, 108219. [CrossRef]
41. Luo, Z.; Sun, M.; Zhang, Z.; Lu, C.; Zhang, G.; Fan, X. Finite element analysis of circle-to-rectangle roll forming of thick-walled rectangular tubes with small rounded corners. *Int. J. Mater. Form.* **2022**, *15*, 73. [CrossRef]
42. Abeyrathna, B.; Ghanei, S.; Rolfe, B.; Taube, R.; Weiss, M. Optimising part quality in the flexible roll forming of an automotive component. *Int. J. Adv. Manuf. Tech.* **2022**, *118*, 3361–3373. [CrossRef]

43. Pacana, J.; Siwiec, D.; Pacana, A. Numerical Analysis of the Kinematic Accuracy of the Hermetic Harmonic Drive in Spac Vehicles. *Appl. Sci.* **2023**, *13*, 1694. [CrossRef]
44. Yan, X.; Liu, Z.; Zhao, Y.; Li, Y.; Chen, W.; Li, M. Experimental and Numerical Study on the Effect of Load Direction on the Bol Loosening Failure. *Eng. Fail. Anal.* **2024**, *163*, 108574. [CrossRef]
45. Yang, F.; Deng, X.; Wang, C. Analysis of energy absorption characteristics of a sinusoidal corrugated filled tube. *Mech. Adv. Mate Struct.* **2024**, *31*, 7386–7403. [CrossRef]
46. Rojas-Sola, J.I.; Sánchez-García, S. Mechanical Study of a Single-Cylinder High-Pressure Steam Engine with a Corliss Valve Gea Using Finite Element Method. *Appl. Sci.* **2025**, *15*, 4782. [CrossRef]
47. Sadighi, M.; Hosseini, S.A. Finite Element Simulation and Experimental Study on Mechanical Behavior of 3D Woven Glass Fibe Composite Sandwich Panels. *Compos. Part B Eng.* **2013**, *55*, 158–166. [CrossRef]
48. Sabuncuoglu, B.; Acar, M.; Silberschmidt, V.V. Finite Element Modelling of Fibrous Networks: Analysis of Strain Distribution ir Fibres under Tensile Load. *Comput. Mater. Sci.* **2013**, *79*, 143–158. [CrossRef]
49. Zimmermann, J.; Stommel, M. The Mechanical Behaviour of Rubber under Hydrostatic Compression and the Effect on the Result of Finite Element Analyses. *Arch. Appl. Mech.* **2013**, *83*, 293–302. [CrossRef]
50. Saxena, A.; Dwivedi, S.P.; Kumaraswamy, A.; Srivastava, A.K.; Maurya, N.K. Influence of SD Effect on Johnson–Cook Hard ening Constitutive Material Model: Numerical and Experimental Investigation for Armor Steel. *Mech. Adv. Mater. Struct* **2020**, *29*, 285–302. [CrossRef]
51. Kweon, H.D.; Kim, J.W.; Song, O.; Oh, D. Determination of true stress–strain curve of type 304 and 316 stainless steels using a typical tensile test and finite element analysis. *Nucl. Eng. Technol.* **2021**, *53*, 647–656. [CrossRef]
52. Harris, A.; Lee, H.P. Finite element analysis and experimental investigation on the mechanical behaviours of multifunctiona sandwich structures embedded with batteries. *Adv. Compos. Mater.* **2021**, *31*, 86–101. [CrossRef]
53. Sudhagar, P.E.; Gunasegeran, M. Evaluation, optimization and prediction of the transverse shear modulus of biomimetic 3D printed sandwich core. *Mech. Adv. Mater. Struct.* **2022**, *31*, 854–868. [CrossRef]
54. Kladovasilakis, N.; Tsongas, K.; Kostavelis, I.; Tzovaras, D.; Tzetzis, D. Effective Mechanical Properties of Additive Manufacturec Triply Periodic Minimal Surfaces: Experimental and Finite Element Study. *Int. J. Adv. Manuf. Technol.* **2022**, *121*, 7169–7189 [CrossRef]
55. Lu, J.-F.; Cheng, J.; Feng, Q.-S. Plane wave finite element model for the 2-D phononic crystal under force loadings. *Eur. J. Mech A-Solids* **2022**, *91*, 104426. [CrossRef]
56. Laly, Z.; Mechefske, C.; Ghinet, S.; Kone, C.T.; Atalla, N. Design and analysis of periodic acoustic metamaterial sound insulator using finite element method. *Noise Control Eng. J.* **2023**, *71*, 344–364. [CrossRef]
57. Danielli, F.; Ciriello, L.; La Barbera, L.; Rodriguez Matas, J.F.; Pennati, G. On the need of a scale-dependent material characteriza tion to describe the mechanical behavior of 3D printed Ti6Al4V custom prostheses using finite element models. *J. Mech. Behav Biomed. Mater.* **2023**, *140*, 105707. [CrossRef] [PubMed]
58. Tonder, J.D.V.; Venter, M.P.; Venter, G. A New Method for Improving Inverse Finite Element Method Material Characterization for the Mooney–Rivlin Material Model through Constrained Optimization. *Math. Comput. Appl.* **2023**, *28*, 78. [CrossRef]
59. Koppisetti, S.B.; Nallu, R. Material characterization of A356-T6 aluminum alloy and correlation of wheel impact test results using finite element analysis. *Int. J. Comp. Mat. Sci. Eng.* **2024**, *13*, 2350025. [CrossRef]
60. Trindade, D.; Habiba, R.; Fernandes, C.; Costa, A.; Silva, R.; Alves, N.; Martins, R.; Malça, C.; Branco, R.; Moura, C. Material Performance Evaluation for Customized Orthoses: Compression, Flexural, and Tensile Tests Combined with Finite Element Analysis. *Polymers* **2024**, *16*, 2553. [CrossRef]
61. Kaselouris, E.; Nikolos, I.K.; Orphanos, Y.; Bakarezos, M.; Papadogiannis, N.A.; Tatarakis, M.; Dimitriou, V. Elastoplastic study of nanosecond-pulsed laser interaction with metallic films using 3D multiphysics fem modeling. *Int. J. Damage Mech.* **2016**, *25*, 42–55 [CrossRef]
62. Kumar, K.K.; Samuel, G.L.; Shunmugam, M.S. Theoretical and experimental investigations of ultra-short pulse laser interaction on Ti6Al4V alloy. *J. Mater. Process. Technol.* **2019**, *263*, 266–275. [CrossRef]
63. Abali, B.E.; Zohdi, T.I. Multiphysics computation of thermal tissue damage as a consequence of electric power absorption. *Comput. Mech.* **2020**, *65*, 149–158. [CrossRef]
64. Sakaev, I.; Linden, J.; Ichaaya, A.A. Dynamic fracture of SiO_2 films due to laser-induced confined micro-explosion at the Si/SiO_2 interface: Time-resolved imaging and finite-element simulation. *Opt. Laser Technol.* **2022**, *150*, 107938. [CrossRef]
65. Wang, Z.; Min, J.; Hu, J.; Wang, Z.; Chen, X.; Tang, Z.; Liu, S. Femtosecond laser–acoustic modeling and simulation for AlCu nanofilm nondestructive testing. *Front. Mech. Eng.* **2024**, *19*, 33. [CrossRef]
66. Smart, T.J.; Abali, B.E.; Boschker, H.; Braun, W. Deposition rates in thermal laser epitaxy: Simulation and experiment. *J. Phys. D Appl. Phys.* **2025**, *58*, 205303. [CrossRef]
67. Bellot, J.-P.; Déculot, L.; Jardy, A.; Hans, S.; Doridot, E.; Delfosse, J.; McDonald, N. Numerical Simulation of the Plasma Arc Melting Cold Hearth Refining Process (PAMCHR). *Metall. Mater. Trans. B* **2020**, *51*, 1329–1338. [CrossRef]

3. Kumar, N.; Shukla, A.; Bansal, S.; Khatri, C.B.; Thakre, G.D.; Yadav, S.K.; Sharma, S.C.; Harsha, S.P. Thermo-hydrodynamic simulation study of twin-groove elliptical (two-lobe) journal bearing of steam turbine with experimental investigations. *Proc. Inst. Mech. Eng. Part J J. Eng. Tribol.* **2021**, *235*, 1745–1764. [CrossRef]

9. Youcef, S.; Mokhefi, A.; Bouzit, M. Theoretical study of thermo-hydrodynamic performances inside an agitated tank equipped with modified two-blade design. *Therm. Sci. Eng. Prog.* **2024**, *55*, 102866. [CrossRef]

). Pegios, I.P.; Hatzigeorgiou, G.D. Finite element free and forced vibration analysis of gradient elastic beam structures. *Acta Mech.* **2018**, *229*, 4817–4830. [CrossRef]

1. Zangeneh, S.; Lashgari, H.R.; Sharifi, H.R. Fitness-for-service assessment and failure analysis of AISI 304 demineralized-water (DM) pipeline weld crack. *Eng. Fail. Anal.* **2020**, *107*, 104210. [CrossRef]

2. Fuad, M.F.I.A.; Mustafa, M.S.; Nazari, A.D.Z.A.; Rashid, Z.A. Gas Piping System Fatigue Life Estimation through Acoustic Induced Vibration (AIV) Analysis. *Int. J. Mech. Eng. Robot. Res.* **2021**, *10*, 276–282. [CrossRef]

3. Liao, J.; Yu, Y.; Yan, J.; Hsu, S.-H. Simulation and modal analysis of marine diesel engine based on finite element model and vibration sensor data. *Sens. Mater.* **2021**, *33*, 1645–1656. [CrossRef]

4. Codina, R.; Türk, Ö. Modal analysis of elastic vibrations of incompressible materials using a pressure-stabilized finite element method. *Finite Elem. Anal. Des.* **2022**, *206*, 103760. [CrossRef]

5. Kaselouris, E.; Alexandraki, C.; Bakarezos, M.; Tatarakis, M.; Papadogiannis, N.A.; Dimitriou, V. A detailed FEM study on the vibro-acoustic behaviour of crash and splash musical cymbals. *Int. J. Circuits Syst. Signal Process.* **2022**, *16*, 948–955. [CrossRef]

6. Sha, Y.; Zhao, W.; Tang, X.; Zhao, F. Acoustic and Vibration Response and Fatigue Life Analysis of Thin-Walled Connection Structures under Heat Flow Conditions. *Aerospace* **2024**, *11*, 287. [CrossRef]

7. Luo, B.; Gallas, S.; Micallef, C.; Govaerts, J.; Gryllias, K.; Poortmans, J. Experimental and finite element modal analysis of photovoltaic modules for the design of next-generation vehicle-integrated PV applications. *Sol. Energy Mater. Sol. Cells* **2025**, *290*, 113683. [CrossRef]

8. Bruchen, J.; Liu, Y.; Moulin, N. Finite element setting for fluid flow simulations with natural enforcement of the triple junction equilibrium. *Comput. Fluids* **2018**, *171*, 103–121. [CrossRef]

9. Alsabery, A.I.; Sheremet, M.A.; Ghalambaz, M.; Chamkha, A.J.; Hashim, I. Fluid-structure interaction in natural convection heat transfer in an oblique cavity with a flexible oscillating fin and partial heating. *Appl. Therm. Eng.* **2018**, *145*, 80–97. [CrossRef]

0. Long, T.; Yang, P.; Liu, M. A novel coupling approach of smoothed finite element method with SPH for thermal fluid structure interaction problems. *Int. J. Mech. Sci.* **2020**, *174*, 105558. [CrossRef]

1. Arif, M.S.; Shatanawi, W.; Nawaz, Y. Finite Element Study of Electrical MHD Williamson Nanofluid Flow under the Effects of Frictional Heating in the View of Viscous Dissipation. *Energies* **2023**, *16*, 2778. [CrossRef]

2. Du, L.; Liang, W.-G.; Xu, G.-G.; Qin, Q.-H.; Li, D.-S. Finite Element Analysis and Computational Fluid Dynamics Verification of Molten Pool Characteristics During Selective Laser Melting of Ti-6Al-4V Plates. *3D Print. Addit. Manuf.* **2023**, *10*, 711–722. [CrossRef] [PubMed]

3. Weiss, A.G.; Zaidan, L.J.; Ahmad, M.T.H.B.; Abdoelatef, M.G.; Peterson, J.W.; Lindsay, A.D.; Kong, F.; Ahmed, K.; Kimber, M.L. Characterization of the Finite Element Computational Fluid Dynamics Capabilities in the Multiphysics Object Oriented Simulation Environment. *ASME J. Nucl. Rad. Sci.* **2023**, *9*, 021402. [CrossRef]

4. Esposito, G.; Tsamopoulos, J.; Villone, M.M.; D'Avino, G. Numerical simulations of the pressure-driven flow of pairs of rigid spheres in elastoviscoplastic fluids. *J. Fluid Mech.* **2025**, *1011*, A20. [CrossRef]

5. Song, J.; Lan, J.; Zhu, L.; Jiang, Z.; Zhang, Z.; Han, J.; Ma, C. Finite Element Simulation and Microstructural Analysis of Roll Forming for DP590 High-Strength Dual-Phase Steel Wheel Rims. *Materials* **2024**, *17*, 3795. [CrossRef]

6. Zou, L.; Wang, X.; Wang, R.; Huang, X.; Li, M.; Li, S.; Jiang, Z.; Yin, W. Crashworthiness Performance and Multi-Objective Optimization of Bi-Directional Corrugated Tubes under Quasi-Static Axial Crushing. *Materials* **2024**, *17*, 3958. [CrossRef]

7. Wang, S.; Wang, B.; Mu, S.; Zhang, J.; Zhang, Y.; Gong, X. Study on the Approach to Obtaining Mechanical Properties Using Digital Image Correlation Technology. *Materials* **2025**, *18*, 1875. [CrossRef]

8. Rodella, A. Extreme Behaviors in Fibrous Material Remodeling: Auxetic to Non-Auxetic Transition and Phase Segregation. *Materials* **2025**, *18*, 1674. [CrossRef]

9. Vasile, A.; Constantinescu, D.M.; Indreș, A.I.; Coropețchi, I.C.; Sorohan, Ș.; Apostol, D.A. Numerical Simulation of Compressive Testing of Sandwich Structures with Novel Triply Periodic Minimal Surface Cores. *Materials* **2025**, *18*, 260. [CrossRef]

0. Kaniolakis Kaloudis, E.; Kaleris, K.; Aravantinos-Zafiris, N.; Sigalas, M.; Katerelos, D.T.G.; Dimitriou, V.; Bakarezos, M.; Tatarakis, M.; Papadogiannis, N.A. Evaluating the Role of Unit Cell Multiplicity in the Acoustic Response of Phononic Crystals Using Laser-Plasma Sound Sources. *Materials* **2025**, *18*, 1251. [CrossRef]

1. Xiong, X.; Wang, W.; Li, X.; Fan, F.; Zhou, J.; Chen, M. Prediction and Optimization of the Long-Term Fatigue Life of a Composite Hydrogen Storage Vessel Under Random Vibration. *Materials* **2025**, *18*, 712. [CrossRef]

92. Bellot, J.-P.; Ayadh, W.; Kroll-Rabotin, J.-S.; Marin, R.; Delfosse, J.; Cardon, A.; Biagi, A.; Hans, S. Experimental and Numerica Study of the Plasma Arc Melting of Titanium Alloys: Application to the Removal of High Density Inclusions (HDIs). *Material* **2025**, *18*, 2051. [CrossRef] [PubMed]
93. Wang, C.; Hua, M.; Wang, L.; Wang, S.; Liu, J.; Liu, R.; Tian, X.; Lin, X. Theoretical and Experimental Research on the Short-Rang Structure in Gallium Melts Based on the Wulff Cluster Model. *Materials* **2025**, *18*, 133. [CrossRef] [PubMed]
94. Huang, L.; Peng, B.; Yue, Q.; Huang, G.; Wang, C.; Wang, R.; Tian, N. First-Principles Study on the Electrical and Therma Conductivities of Cu–Zn Binary Alloys. *Materials* **2025**, *18*, 2310. [CrossRef] [PubMed]

materials

MDPI

Article

Finite Element Simulation and Microstructural Analysis of Roll Forming for DP590 High-Strength Dual-Phase Steel Wheel Rims

ingwen Song [1,†], Jun Lan [1,†], Lisong Zhu [2], Zhengyi Jiang [2], Zhiqiang Zhang [3], Jian Han [1,*] and Cheng Ma [4,*]

1 School of Materials Science and Engineering, Tianjin University of Technology, Tianjin 300384, China; sjw13314701971@163.com (J.S.); lj526586@163.com (J.L.)
2 School of Mechanical, Materials, Mechatronic and Biomedical Engineering, University of Wollongong, Wollongong, NSW 2522, Australia; lz131@uowmail.edu.au (L.Z.); jiang@uow.edu.au (Z.J.)
3 Han-Steel, Handan 056015, China; zzq198265@163.com
4 Technology Research Institute, HBIS Group, Shijiazhuang 050023, China
* Correspondence: hj_tjut@vip.126.com (J.H.); macheng01@hbisco.com (C.M.)
† These authors contributed equally to this work.

Abstract: In this study, finite element (FE) simulation by the software Abaqus was relied on to investigate the roll forming process of a wheel rim made of an innovative dual-phase steel, i.e., DP590, after flash butt welding (FBW). In the simulation, an FE model was generated, including the design of the dies for flaring, three-roll forming, and expansion, and detailed key processing parameters based on practical production of the selected DP590. Combined with the microstructures and properties of the weld zone (WZ) and heat-affected zones (HAZs) after FBW, the distribution of stress/strain and the change in thickness of the base metal (BM), WZ and HAZs were analyzed, and compared in the important stages of roll forming. Theoretically, the variation in the microstructure and the corresponding stress–strain behaviors of the BM, WZ, and HAZs after FBW have led to the thickness reduction of DP590 that originated from softening behaviors occurring at the region of subcritical HAZs (SCHAZs), and a small amount of tempered martensite has evidently reduced the hardness and strength of the SCHAZ. Meanwhile, the distribution of stress/strain has been influenced to some extent. Further, the study includes the influence of the friction coefficient on the forming quality of the wheel rim to guarantee the simulation accuracy in practical applications. In sum, the dual-phase steel has to be carefully applied to the wheel rim, which needs to experience the processes of FBW and roll forming, focusing on the performance of SCHAZs.

Keywords: flash butt welding; dual-phase steel DP590; FE modeling; roll forming

1. Introduction

As an important component, the wheel rim plays an important role in the smooth and safe operation of automobiles [1]. The rim and the tire support each other and buffer the force transmitted by the ground, so the performance of the entire wheel is largely determined by the performance of the rim. With the rapid development of the manufacturing industry, various research and development institutes have carried out a series of technical studies and research on steel wheels to improve the strength, fatigue, and other mechanical properties of the wheel rim, as well as the comprehensive performance of welding and forming, to ensure the safety of the wheel rim [2,3]. Modern tubeless rims generally use equal-thickness steel plates as raw materials [4], with a 2.5–8 mm thickness range. The general process of their production isinvolves (1) uncoiling, (2) blanking, (3) rolling, (4) flash butt welding, (5) slag scraping, (6) flaring, (7) one-pass rolling, (8) two-pass rolling, (9) three-pass rolling, (10) expansion and finishing, (11) air tightness test, and (12) punching valve holes. The process diagram is shown in Figure 1.

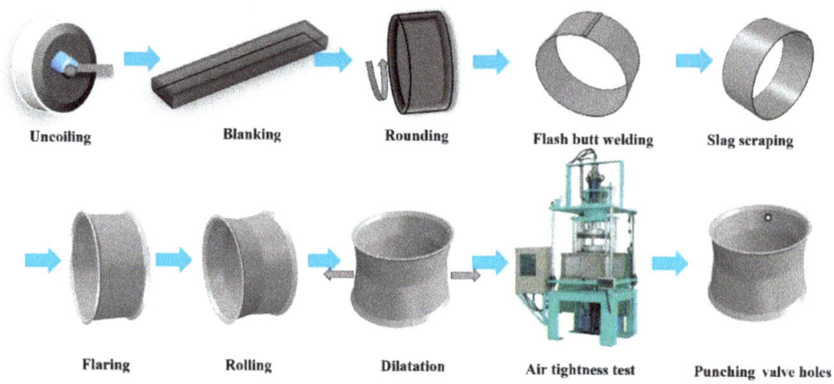

Figure 1. Rim production process diagram.

The rim material is made using high-strength steel DP590. Dual-phase (DP) steel is a low-carbon steel alloy subjected to critical-zone heat treatment or a controlled rolling process to obtain martensite and ferrite at a faster cooling rate [5]. Since the dual-phase steel has a hardened phase distributed on the fine ferrite matrix and continues to be strengthened by solid solution atoms [6], 5–20% martensite guarantees the strength of the material. With the increase in martensite content, the strength of DP steel can reach 500–1200 MPa [7,8].

There is no obvious yield point in the stress–strain curve of DP steel, which effectively avoids the wrinkling problem on the surface of the formed parts. DP steel has a low yield strength and high tensile strength, which makes it have a high yield ratio. The strength of the workpiece during the forming process is low, and the strength after forming is high. Therefore, the formed parts have high impact energy absorption and fatigue strength. During the deformation process, the elongation of DP steel is uniform and the total elongation is large [9]. Therefore, dual-phase steel has good comprehensive mechanical properties and is a potential material for rims [10].

Flash butt welding technology is the mainstream method for rail welding. In addition to being used in the field of rails, it is also used for the closed welding of open blanks and parts, such as automobile rims, bicycle rims, and motorcycle rims [11]. Because it has no need to add other materials [12] and has a high welding efficiency, it has become the mainstream welding method for wheel rim production. In the flash butt welding process, high-temperature plastic deformation occurs in the weld and the heat-affected zone near the weld, which leads to a change in the microstructure and mechanical properties [13], thus affecting the subsequent forming process. Therefore, the roll-forming quality of welded joints is closely related to the welding process parameters [14,15].

Compared with the traditional design method, numerical simulation technology has great advantages in terms of cost, efficiency, and working environment. More and more mechanical manufacturing processes have been unprecedentedly developed with the support of finite element technology [16]. A reasonable finite element model is established to provide scientific researchers and engineers with the dynamic details of the product during processing [11]. The important processes in rim production were studied, and the influence law of key process parameters was found and was reasonably regulated, which can improve product quality, shorten the development cycle, and cut costs [17].

There has been a lot of research on the numerical simulation of the roll forming process, and some experience has been summarized. Kiuchi [18] developed a computer-based numerical simulation system for cold roll forming, which is suitable for various roll forming processes. It is convenient to analyze the influence of rolling pass, roll position, product size, and sheet mechanical properties on forming. Abe et al. [19] used the rigid-plastic finite element method to simulate the multi-pass forming of spokes, discussed the thinning of wall fillet thickness in each stage, and studied the influence of fillet punch radius and tensile ratio on fillet thinning. Bi et al. [20] carried out a numerical simulation of the spinning of automobile rims and improved the spin forming process. The research results have a certain reference value for the study of rim spin forming. Fang et al. [21] established a finite element model of the rolling process in ABAQUS, studied the law of rim roll forming, and analyzed the deformation results of each pass. The reliability of the model was verified, which provides a reference for the numerical simulation of high-precision rims. Finally, the problem of rim cracking was analyzed, and the influence of different flaring die angles on the forming results of the rim was explored. Lu et al. [22] established a simulation model of rim plastic forming, studied some key technologies of the model, analyzed the distribution of stress and strain, and compared it with the actual production of the rim to verify the accuracy of the model. Hwang et al. [23] used an offline hot-rolling simulator and numerical simulation technology to simulate and analyze the measured temperature of each region of the workpiece.

In comparison to existing research, this study focuses on DP590 steel wheel rims and utilizes ABAQUS 2020 finite element analysis software to establish models to investigate the key mechanisms involved in the forming process of wheel rims post-flash welding. The study integrates the processes of flash welding and roll forming, delineating the welding seam and heat-affected zone within the model. Through microscopic structural analysis, it correlates the stress–strain distribution and thickness during roll forming with the actual microstructure, revealing the association between microstructural changes and macroscopic forming quality. The study explores the impact of critical process parameters on this process to ensure the accuracy of simulation results in practical applications. The findings of this research are crucial for optimizing wheel rim manufacturing processes, enhancing product quality, and reducing production costs. They provide new insights into improving wheel rim roll forming processes and serve as a reference for future studies.

2. FEM Model of the Roll Forming of a Wheel Rim

2.1. Die Design

The research object of this study is DP590, and its alloy composition and mechanical properties parameters, obtained by tensile test, are shown in Table 1.

Table 1. DP590 alloy elements and mechanical properties.

Alloying Element	**Elements wt.%**	**C 0.056**	**Mn 1.199**	**Cr 0.268**	**Si 0.085**	**Al 0.039**	**Ti 0.02**	**Nb 0.013**	**P 0.015**	**N 0.004**	**S 0.001**	**Fe Bal.**
Performance parameter	Yield strength						318 MPa					
	Ultimate tensile strength						614 MPa					
	Elongation rate						39.3%					
	Young's modulus						199 GPa					
	Poisson ratio						0.33					

The GB/T 31961-2015 [24] profile of a 15° deep-groove rim of 22.5 × 7.50 truck and bus rims, is shown in Figure 2a. It is welded into a ring by plate winding, flaring at both ends, and then formed by three passes of roll bending. The last step is finishing.

In the simulation process, the mold of the rim-forming process can be designed according to the target aim of segmented deformation and the principle of a constant centerline position before and after deformation. The equal gap theory is used to design the size of each pass of the rolling mold according to the final size diagram, as shown in Figure 2b–f.

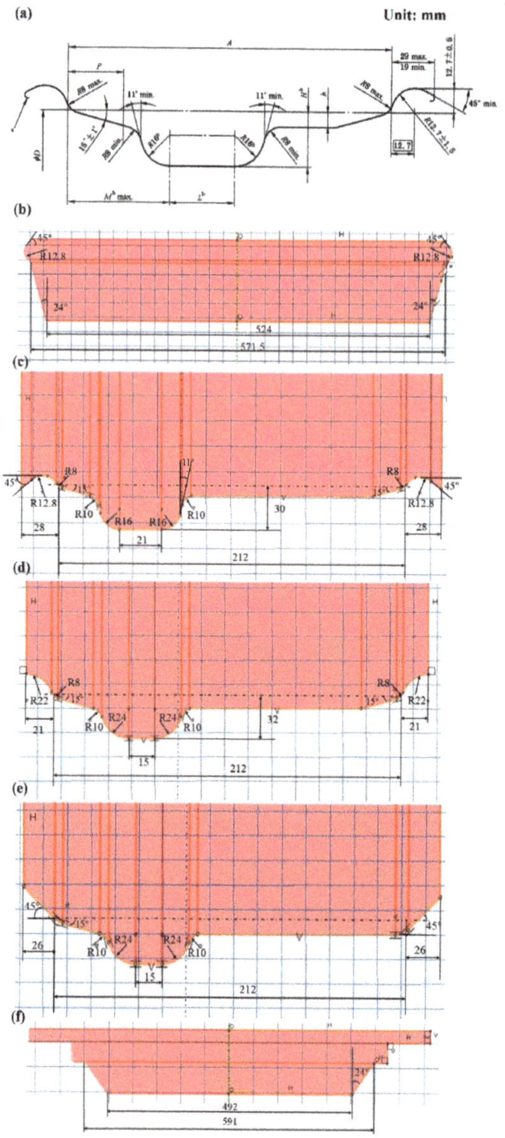

Figure 2. (**a**) 15° deep-groove rim national standard size; mold size: (**b**) expansion die size; (**c**) third-roll die size; (**d**) second-roll die size; (**e**) first-roll die size; (**f**) flaring die size.

2.2. Finite Element Models of the Roll Forming of the Wheel Rim

The commercial finite element software ABAQUS/Explicit 2020 was used as the simulation tool. ABAQUS/Explicit is a finite element software designed for simulating and analyzing complex nonlinear problems, particularly suitable for fast dynamic events and large deformations. Its primary applications include fast dynamic analysis, large deformation problems, complex contact phenomena, and nonlinear material behavior simulation. It effectively handles high-speed dynamic events such as impacts, collisions, and explosions, providing accurate results for metal forming, material fracture, and collapse, while addressing issues like contact separation, sliding, and friction [25]. It is widely used in the automotive and aerospace industries and in civil engineering [26], manufacturing, and biomedical engineering [27]. However, ABAQUS/Explicit requires significant computational resources during explicit simulations, especially with high-resolution models and long-duration simulations. The choice of time step directly impacts the stability and accuracy of the simulation results. To ensure accuracy, it is common practice to validate and calibrate simulation models using experimental data, and to conduct error analysis to understand potential sources of numerical errors. Through the optimization of parameters and model settings, ABAQUS/Explicit can provide simulation outputs highly consistent with experimental results, enabling in-depth exploration of the nature and principles behind complex dynamic phenomena.

Figure 3 shows the model for roll forming. The forming system includes an upper die, lower die, and guide wheel. The upper die and the lower die have convex and concave profiles, respectively. Initially, the flaring ring blank is placed on the lower die. Once the model is started, the upper die and the lower die begin to rotate. The lower die moves upward to contact the upper die, and the annular blank rotates at the same speed as the lower die during the clamping process. The guide wheel and retaining ring are used to stabilize the blank and prevent it from shaking.

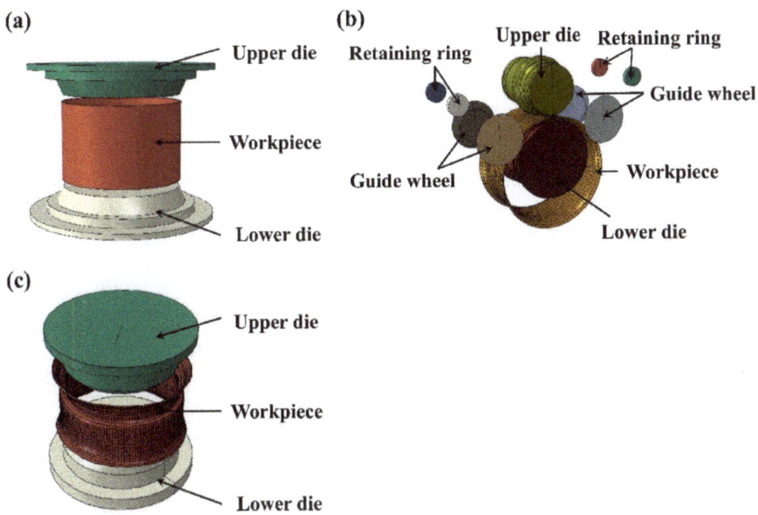

Figure 3. Finite element model of each process of rim production: (**a**) flaring; (**b**) rolling; (**c**) expansion.

To explore the stress–strain law of the roll forming process of the welded rim, the weld and the heat-affected zone are separately divided when the model is established. Based on the peak temperature of each region, the HAZ can be divided into four main regions: the coarse-grained heat-affected zone HAZ1 (CGHAZ), the fine-grained heat-affected zone HAZ2 (FGHAZ), the inter-critical heat-affected zone HAZ3 (ICHAZ), and the subcritical heat-affected zone HAZ4 (SCHAZ). According to their microstructure and properties, different material parameters are assigned to each region, and a width of 3 mm is set, respectively [28], as shown in Figure 4. The true stress–true strain parameters of each region are introduced after conversion, and the changes in stress–strain and thickness reduction near the weld are observed.

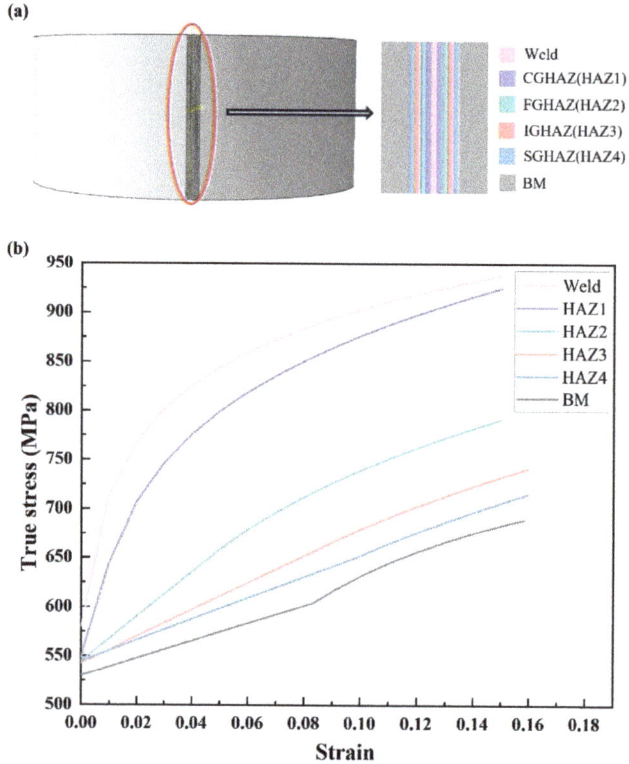

Figure 4. (**a**) Heat-affected zone area division diagram; (**b**) true stress–true strain curves of BM, WZ, and HAZ.

In the process of rim roll forming, three-dimensional finite element modeling is used. Since die deformation during the rolling process can be neglected, the die is modeled as an analytical rigid body. To consider the change in the thickness of the workpiece during the rolling process, the radial force and the axial force are analyzed. In this study, the SC8R solid-shell element was used for meshing, and a total of 4528 elements and 9362 nodes were generated. To ensure the reliability and accuracy of the computational results, we conducted a detailed mesh convergence analysis. We progressively refined the mesh, doubling the number of elements each time, and selected the stress, strain, and deformation at critical locations as the convergence criteria. Additionally, we analyzed the reaction forces and displacements at key points during the rolling process. By comparing the results at different mesh densities, we found that when the number of elements increased to 4528, the variation rate of the stress and deformation at critical locations stabilized, with a change rate of less than 1%, meeting ABAQUS's convergence criteria. We finally selected a mesh with 4528 elements and 9362 nodes, achieving a good balance between computational efficiency and accuracy. Further mesh refinement showed that the variation rate of the results at critical locations remained below 1%, indicating that the mesh had converged.

In this study, the metal roll forming simulation calculation is carried out using the finite element method. Considering various boundary conditions, such as mold shape, friction, rotation speed, etc., a three-dimensional model is established for nonlinear calculation, and the stress–strain, thickness, material flow, and other related information in the forming process are obtained. The elastic-plastic finite element method takes into account the elastic deformation of the material. It can analyze the dynamic motion of the plastic deformation zone, the stress and strain changes during processing and forming, and can better simulate the actual production and processing process.

Several basic rules of the elastic-plastic finite element method are as follows:

Yield criterion:

When the equivalent stress exceeds the yield stress of the material, the material enters the plastic deformation stage, which satisfies the condition of the yield criterion:

$$F^0 = F^0(\sigma_{ij}, k_0) = 0 \tag{1}$$

In the formula, σ_{ij} is the stress tensor component, k_0 is the material characteristic parameter, and F^0 is the initial yield surface.

When the metal material is deformed, the commonly used yield conditions are:

(a) Mises condition:

$$F^0(\sigma_{ij}, k_0) = F(\sigma_{ij}) - k_0 = 0 \tag{2}$$

Among which,

$$f(\sigma_{ij}) - k_0 = \frac{1}{2} s_{ij} s_{ij}\, ;\, k_0 = \frac{1}{3}\sigma_{s0}^2 \tag{3}$$

$$s_{ij} = \sigma_{ij} - \sigma_m \delta_{ij}\, ;\, \sigma_m = \frac{1}{3}(\sigma_{11} + \sigma_{22} + \sigma_{33}) \tag{4}$$

In the formula, f is the plastic stress–strain function, s_{ij} is the stress component, σ_{s0} is the initial yield stress of the material, and σ_m is the normal stress.

The relationship between s_{ij} and equivalent stress:

$$\frac{1}{2} s_{ij} s_{ij} = \frac{\overline{\sigma}^2}{3} \tag{5}$$

(b) Tresca condition:

In the three-dimensional principal stress space, the yield surface is a regular hexagonal prism with $\sigma_1 = \sigma_2 = \sigma_3$ as the axis and an inscribed Mises cylindrical surface. Its yield trajectory on the π plane is a regular hexagon with an inscribed Mises yield trajectory. The yield function expression is:

$$F^0(\sigma_{ij}, k_0) = \frac{1}{6}\left[(\sigma_1 - \sigma_2)^2 + (\sigma_2 - \sigma_3)^2 + (\sigma_3 - \sigma_1)^2\right] - \frac{1}{3}\sigma_{s0}^2 \tag{6}$$

Flowing rule:

The relationship between plastic strain increment and the stress state is expressed by Prandt-Reuss:

$$d\varepsilon_{ij}^p = d\lambda \frac{\partial f}{\partial \sigma_{ij}} \tag{7}$$

In the formula, $d\varepsilon_{ij}^p$ is the plastic strain increment component, f is the plastic stress–strain function, and λ is the proportional constant. this Equation indicates that $d\varepsilon_{ij}^p$ is perpendicular to the surface defined by f = 0.

Hardening criterion:

The hardening criterion of the material indicates that the initial yield criterion changes with the increase in plastic deformation. The hardening criterion specifies the material loading function under plastic deformation:

$$F(\sigma_{ij}, k, a_{ij}) = 0 \tag{8}$$

In the formula, σ_{ij} is the stress tensor component, k is the plastic work, and a_{ij} is the moving tensor of the loading surface.

In actual rim processing, to prevent the rim from bouncing or offsetting when it contacts the upper and lower rolling molds, guide wheels are set on both sides of the rim.

The guide wheel is mainly used to transfer the pressure provided by the cylinder to the rim. Due to the complex contact between the guide wheel and the edge of the rim, the model of the guide wheel is simplified in the process of numerical simulation, and the force provided by the cylinder is replaced by establishing a spring and adjusting the elastic coefficient and the damping coefficient. According to the calculation formula of elastic coefficient k_{max}, the elastic coefficient is set to 1000 N/mm, and the damping coefficient is set to 0.3.

In the flaring process, the upper and lower dies are in contact with the inner surface of the workpiece, respectively. In the three-pass rolling process, the upper die is in contact with the outer surface of the workpiece, the lower die is in contact with the inner surface of the workpiece, and the side guide wheel is in contact with the side of the workpiece. In the numerical simulation, the contact is surface-to-surface contact, and the Coulomb friction model of the penalty function is used. In friction contact, the tangential direction is the friction behavior, the friction coefficient is defined as 0.3, and the normal direction adopts the default 'hard' contact. In the simulation process, the rotation speed of the upper rolling die is 180 r/min, the rotation speed of the lower rolling die is 216 r/min, and the feed rate of the lower rolling die is 15 mm/s.

3. Results and Discussion

3.1. Microstructure and Properties

Figure 5 shows the microstructure of different heat-affected zones at the weld. The microstructure of the DP590 steel matrix metal is composed of 12.81% martensite (M) and 87.19% polygonal ferrite (PF). The WZ has coarse grain and Widmanstatten structure [29]. Using the CALPHAD software "Thermal-Calc" (Thermo-Calc Software, Stockholm, Sweden), the equilibrium phase diagram of DP590 was calculated. The results show that the Ac_1 and Ac_3 temperatures are 665 °C and 844 °C, respectively. During the flash butt welding process, the heat-affected zone is divided according to Ac_1 and Ac_3: the peak temperatures of the CGHAZ and FGHAZ regions are higher than Ac_3, leading to complete recrystallization. The grain size in CGHAZ is generally greater than 100 μm, while in FGHAZ it is mostly less than 50 μm. The peak temperature in the ICHAZ is between Ac_1 and Ac_3, resulting in partial recrystallization. The peak temperature in the SCHAZ is lower than Ac_1 but higher than 200 °C, so no austenite transformation occurs, but it still affects performance.

In Figure 5c,d, the CGHAZ and FGHAZ show a lath martensite structure and austenite grain boundaries, with the CGHAZ having larger original austenite grains than the FGHAZ. The ICHAZ consists mainly of fine ferrite and some retained tempered sorbite. The SCHAZ's microstructure is similar to that of the base metal, but with larger ferrite grains and a lower martensite content.

The average microhardness values for different regions of the weld are: CGHAZ (232.7 $HV_{0.5}$) > FGHAZ (225.8 $HV_{0.5}$) > ICHAZ (187.7 $HV_{0.5}$) > BM (184.2 $HV_{0.5}$) > SCHAZ (167.2 $HV_{0.5}$) [29]. The microhardness of the ICHAZ and SCHAZ is significantly lower than that of the coarse-grained and fine-grained regions. In terms of softening behavior, the hardness of the SCHAZ decreased by 9.2% compared to the BM. The welded heat-affected zone of DP590 steel is defined as an abnormal softening zone that must meet usage requirements. Figure 5 shows that the CGHAZ and FGHAZ have a lath martensite structure, providing high hardness. In contrast, the hardness of the ICHAZ is significantly reduced, mainly due to fine ferrite grains and a small amount of tempered martensite. The microstructure of the SCHAZ is very similar to that of the BM, but its martensite exhibits characteristic tempering, which is the cause of softening in the SCHAZ.

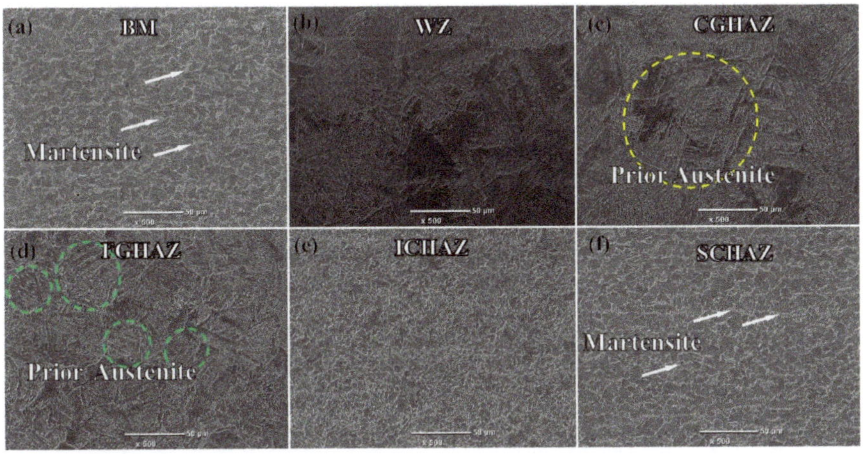

Figure 5. Microstructure of the main heat-affected zones. (**a**) BM; (**b**) WZ; (**c**) CGHAZ; (**d**) FGHAZ; (**e**) ICHAZ; (**f**) SCHAZ.

3.2. Stress and Strain in the Forming Process

The distribution of Von Mises stress (S, Mises) and equivalent plastic strain (PEEQ) of the flaring blank is shown in Figure 6. The results indicate that the primary equivalent stress in the workpiece is 372.2 MPa, while the maximum equivalent stress at the weld reaches 553.5 MPa. The majority of deformation during the flaring process occurs at the rim area. After flaring, the rims on both sides of the workpiece take on a horn-like shape, with strain distributed symmetrically. The peak strain observed is 6.07%.

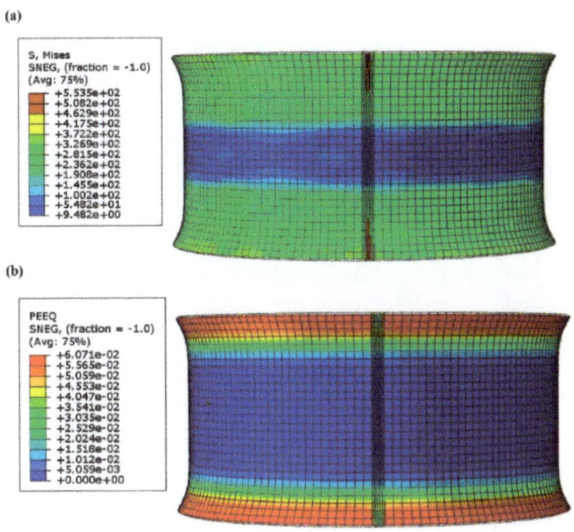

Figure 6. (**a**) Stress distribution mapping of flaring process; (**b**) strain distribution mapping of flaring process.

The distribution of the equivalent stress and plastic strain of the flaring ring formed by the first roll bending is shown in Figure 7(a_1,b_1). The weld area experiences stress concentration due to the layered martensitic structure, with the maximum stress reaching 689.4 MPa, primarily localized on both sides of the groove. The coarse grain structure in the CGHAZ and FGHAZ also leads to higher stress during the forming process [30]. The key forming sections in the rolling process are the middle groove and the flanges on either side. The groove is pre-formed, and the flanges are shaped further based on the horn-like form created during flaring. The deformation at the rounded corners on both sides of the groove is the most significant, with a maximum strain of 4.2%, followed by other parts of the groove. The rolling process involves the most extensive deformation in the three-pass process. The plastic strain at the left corner of the groove is greater than at the right corner because, during rolling, material primarily flows from the left side of the groove to the groove's base, leading to a tendency toward thinning.

The stress and strain distribution of the workpiece during the second rolling process is shown in Figure 7(a_2,b_2). Compared with the first pass, the primary goal of the second rolling process is to refine the shape: the groove bottom remains unchanged while the rim is further shaped. In the second pass, the workpiece is mainly subjected to an equivalent stress of about 454.2 MPa. Stress concentration occurs at the fillet weld near the flange side of the groove, which is related to the softening in the ICHAZ and the SCHAZ. These areas are structurally weaker and more prone to thickness reduction. Additionally, the simulation results show significant deformation at the rim near the groove, indicating high radial and axial stress. The maximum stress is 748.6 MPa, correlating with the presence of larger ferrite grains and layered martensite in the microstructure [31]. Regarding the PEEQ, the maximum strain, about 3.77% is found at the groove. The middle groove continues to exhibit substantial strain, with the strain at the left corner of the lower groove exceeding that of the right corner.

After two passes of roll forming, the rim contour is almost formed, except for the corner of the groove. The purpose of the third pass of roll forming is to leave each fillet of the groove and the rim completely formed, and the feed rate is increased compared with the first two processes. Figure 7(a_3,b_3) shows the equivalent stress and plastic strain on the forming contour after the third pass of roll forming. In the three-roll process, the rim is first deformed, and then the fillets of each part of the groove is also deformed to varying degrees. The stress concentration occurs at the fillet welds on both sides of the groove, and the maximum stress here is 697.2 MPa. This may be because the material flow mainly occurs in these areas during the three rolling processes, which is consistent with the microstructure in the CGHAZ and FGHAZ, indicating that the layered martensite with a high hardness may be subjected to great stress during the forming process. The maximum PEEQ appears at the rim edge with the thickest contour wall, and the strain is 3.64%.

As shown in Figure 8, the stress and strain distribution of the workpiece after expansion is also called bulging. The expansion process is designed to expand the diameter of the workpiece to the design requirements through the expansion die so that the workpiece flattens. The stress value in the larger stress area is about 558.6 MPa. The strain at the groove is the largest, about 2.33%. In the process of expansion, the diameter of the groove is the smallest; first, contacting the mold and driving the whole to expand. After measurement, the diameter before expansion is 565.8 mm, the diameter after expansion is 571.5 mm, and the expansion rate is about 1%.

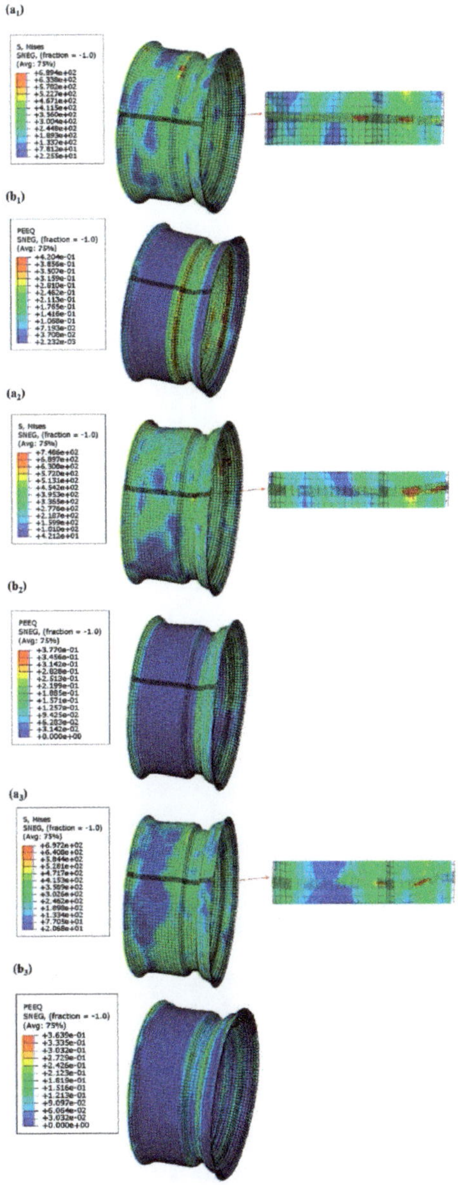

Figure 7. The stress and strain distribution maps of the rolling process: (**a**) stress distribution map, (**b**) strain distribution map; ($\mathbf{a_1}$,$\mathbf{b_1}$) the first pass of rolling, ($\mathbf{a_2}$,$\mathbf{b_2}$) the second pass of rolling, ($\mathbf{a_3}$,$\mathbf{b_3}$) the third pass of rolling.

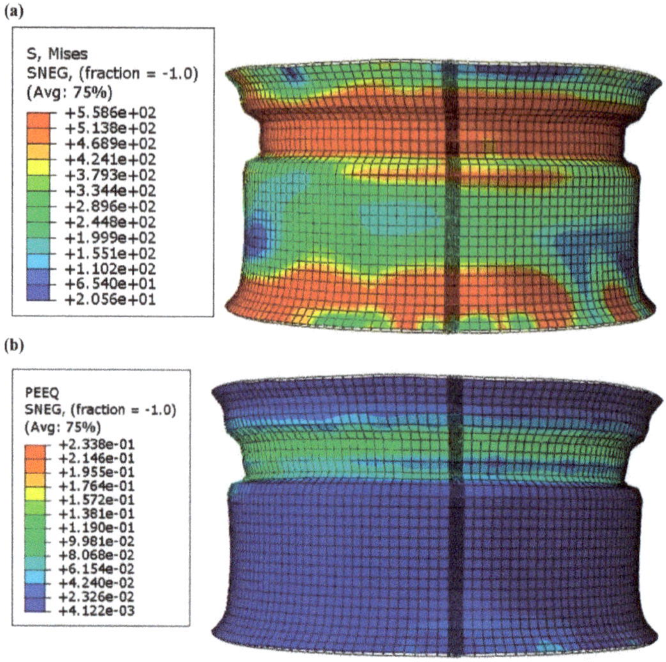

Figure 8. (**a**) Stress distribution mapping of expansion process; (**b**) strain distribution mapping of expansion process.

3.3. Thickness Analysis in Rim-Forming Process

Along the path from the left rim to the right rim, the thickness variation curve of the three-pass roll forming process of the base metal, weld, and heat-affected zone of the rim is output, and the thinning of each region during the rim-forming process is explored.

The thickness variation curve of different regions of the rim after the first roll is shown in Figure 9a. The main forming parts include the groove and the flanges on both sides. On the left flange, the base material experiences the greatest thinning, with a reduction rate of 3.06%. Secondly, the heat-affected zone HAZ4 has a thinning rate of 2.77%, likely due to the hardness reduction caused by carbon diffusion, martensite decomposition, and tempering of the material. This softening effect may lead to a decrease in hardness in this area, ultimately resulting in thickness reduction. The thickness changes in the weld and HAZ1 are similar, indicating signs of material hardening in these areas. This hardening is primarily due to stress concentration during the first rolling process, which induces work hardening in the material [32]. Since the martensitic structure in the CGHAZ and FGHAZ can provide higher hardness, the rate of thickness reduction in these regions is relatively low. However, the groove exhibits the highest thinning rate, reaching 3.83%, likely due to stress concentration in this area during the first rolling process, resulting in material stretching and deformation.

Figure 9b shows the thickness variation curve of different regions of the rim after the two-roll process. After the second rolling, both the flanges on either side of the rim and the groove experience thinning. On the right side of the groove, near the flange, the thinning is particularly noticeable. This is primarily because, during the second rolling process, the flange and the groove bottom deform almost simultaneously, causing further forming of the flanges on both sides. The right side of the groove is subjected to greater tensile stress, with a thinning rate of 7.33%. This high thinning rate phenomenon may be related to the softening of the structure in the FGHAZ and ICHAZ, making these areas more prone to thickness reduction and stress concentration during the forming process. The thinning pattern in the various regions of the flanges on both sides remains consistent with that of the first rolling. The base material on the left flange experiences the greatest thinning, with a reduction rate of 3.05%, followed by HAZ4 with a reduction rate of 2.77%. However, the differences in thinning across the various regions of the flanges are less pronounced than in the first rolling. The differences in thinning rates in the various regions of the groove are not significant, averaging around 9.67%.

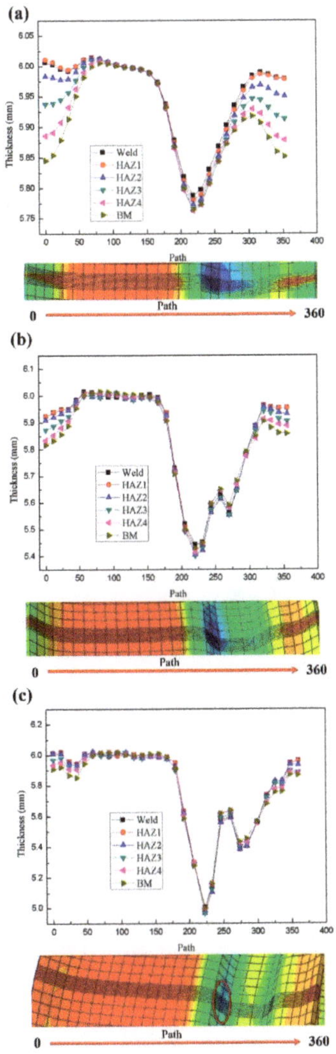

Figure 9. The thickness change curve of different parts of the rim: (**a**) the first rolling; (**b**) the second rolling; (**c**) the third rolling.

According to Figure 9c, the change in rim thickness after three rolls is similar to that after two rolls, but the deformation is the smallest. Due to the superposition of thickness from the three rolling processes, the thinning at the groove is the greatest, with a reduction rate reaching 17%. This may be related to the martensitic structure in the CGHAZ and FGHAZ, as well as the smaller grain size in the FGHAZ. These microstructural characteristics may lead to greater stress in this region during the forming process, resulting in thickness reduction. On the left side of the groove near the rim, significant deformation from the second and third rolling processes results in considerable tension, with a thinning rate of 9.67%. After the third rolling process, ear widths form on both sides of the flanges, and the difference in thinning rate among the various regions is minimal. This is mainly because the shape of the flanges ensures a relatively uniform thickness distribution.

3.4. The Influence of Process Parameters on the Rim-Forming Process

3.4.1. The Influence of Friction Coefficient on the Forming Process

In roll forming, the rim rotates with the upper and lower molds by friction. However, due to the influence of lubrication conditions and mold surface quality, it is difficult to determine the relationship between the friction and the sensitive parts. According to the empirical values in the literature, the effects of friction coefficients of 0.1, 0.2, 0.3, and 0.4 on the forming quality of the rim were investigated in the simulation.

Since the weld seam at the groove of the rim is the most obvious area of thinning, the thickness change in the weld seam after one roll forming pass with different friction coefficients is mainly studied. Figure 10a shows the influence of different friction coefficients on the thickness of one side rim, and the path from left to right is rim-to-groove. It can be seen that when the friction coefficient is 0.1, the flange thickness is smaller and the groove thickness is larger; when the friction coefficient is 0.2, the thickness of the groove is the smallest and the thinning rate of the groove is the highest. The thinning effect of a friction coefficient of 0.3 is similar to that of 0.2. Because groove forming involves material distribution and flow, too small a thickness will lead to an excessive thinning rate, and balanced thickness distribution is more conducive to forming quality. Figure 10b shows the influence of different friction coefficients on the equivalent stress. The higher the friction coefficient, the smaller the equivalent stress and the lower the forming difficulty. Based on the above results, the forming quality is best when the friction coefficient is 0.4.

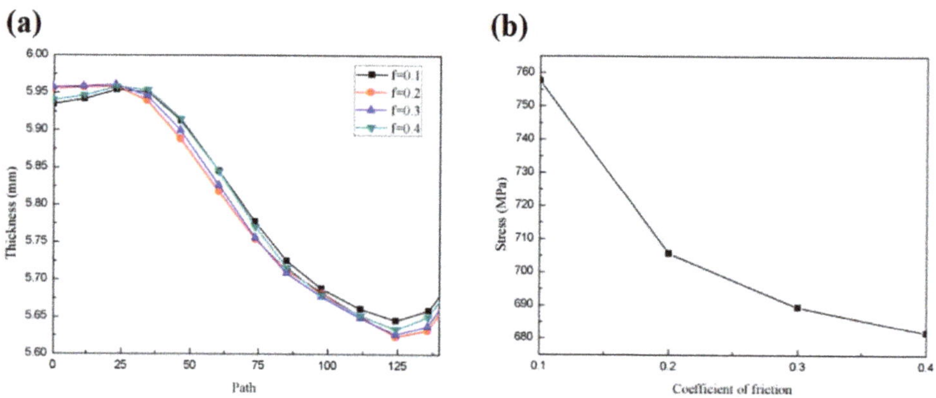

Figure 10. (**a**) Effect of different friction coefficients on rim reduction; (**b**) effect of different friction coefficients on equivalent stress.

3.4.2. The Influence of Feed Rate on the Forming Process

The feed distance refers to the distance during which the lower roll feeds upwards, contacts the inner surface of the workpiece, and continues to drive the workpiece upward until the outer surface of the workpiece contacts the upper roll. If the feed speed is too high, it can easily cause the workpiece to "deviate", resulting in asymmetrical wheel flanges after forming. Additionally, significant material flow occurs in the groove formation during the first rolling process. Once the groove position is determined, it lays the foundation for subsequent processes, making analysis of the quality of the first roll forming critical. The feed speed affects the force exerted on the workpiece by the mold; excessive force increases the difficulty of the rolling process and destabilizes the model.

The variation curve of the maximum support reaction force of different feed speeds obtained by simulation is shown in Figure 11a. The back force represents the interaction force between the workpiece and the guiding wheel, reflecting the stability of the forming process. As the feed rate increases, the model experiences higher back forces, indicating increased difficulty and instability in the forming process. In practical production, prolonged exposure to significant loads and impacts can prematurely fatigue equipment components, reducing equipment stability and accuracy, and potentially causing the wheel rim forming to deviate. Figure 11b illustrates the variation curve of the minimum thickness of the wheel rim at different feed rates. It shows a trend where the minimum thickness initially increases and then decreases with an increasing feed rate. A larger minimum thickness indicates less thinning of the wheel rim during forming, indicating higher forming quality. When the feed rate increases to 30 mm/s, the thinning rate reaches 6.47%, accompanied by higher back forces, which are unfavorable for stable forming. Considering the impact of feed rate on back force and thickness, a feed rate of 15 mm/s provides better stability for the forming model.

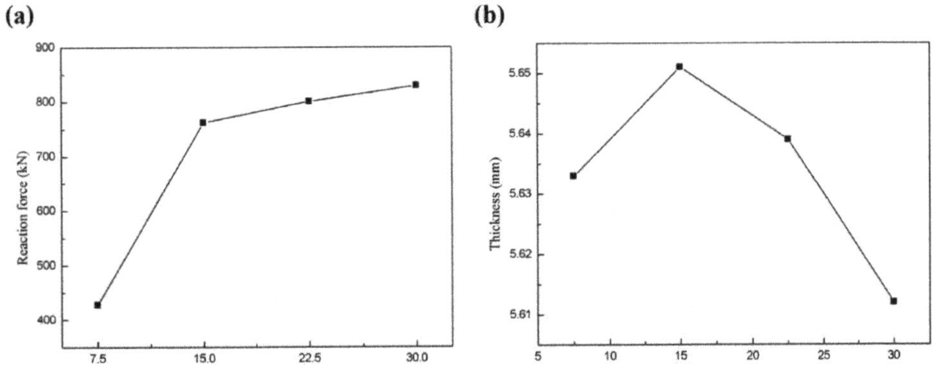

Figure 11. (**a**) Curve of the reaction force at different feed speeds; (**b**) curve of minimum thickness at different feed rates.

3.4.3. The Influence of Mold Speed on the Forming Process

When the rim is rolled, the down roll of the driving wheel drives the workpiece to rotate, and its speed is larger than the up roll, which can make workpiece forming easier. The mold speed affects the initial speed of the rim starting to rotate. Taking 180 r/min as the up-rolling speed, ratios of 1:1, 1:1.2, 1:1.5, and 1:1.8 were set up to study the influence of down-rolling speeds of 180 r/min, 216 r/min, 270 r/min, and 324 r/min on the rim-forming process.

The speed of rolling down will affect the initial speed of rim rotation and the stability of the model. Figure 12 shows the effect of the down rolling speed on the maximum equivalent stress and circumferential stress. The circumferential stress refers to the stress acting on the rim circumference (the direction around the rim). These stresses are caused by the pressure applied in the forming process, the transfer of force, the deformation of the material, and the effect of the forming die. It can be seen from this Figure that the equivalent stress extremum increases with the increase in the rolling speed. The equivalent stress extremum of the one-roll forming process is mainly concentrated near the fillet of the groove. An excessive stress value can easily lead to stress concentration, making it a dangerous part. From the extreme value of circumferential stress, when the rotational speed of the lower die is 180 r/min, the extreme value of circumferential stress is at its largest, reaching 758 MPa.

An excessively low rotational speed increases the difficulty of forming, and excessively high circumferential stress will increase the likelihood of circumferential cracking of the rim. With the increase of rotational speed, the circumferential stress extremum decreases first and then increases. When the rolling speed is 216 r/min, the circumferential stress extremum is at its lowest, which is 692 MPa. When the rotational speed continues to increase, the extreme value of circumferential stress increases again, but the change is small. In order to reduce the difficulty of forming and ensure that the stress on the rim cannot be too great, a ratio of up- and down-rolling speed of 1:1.2 is the most suitable.

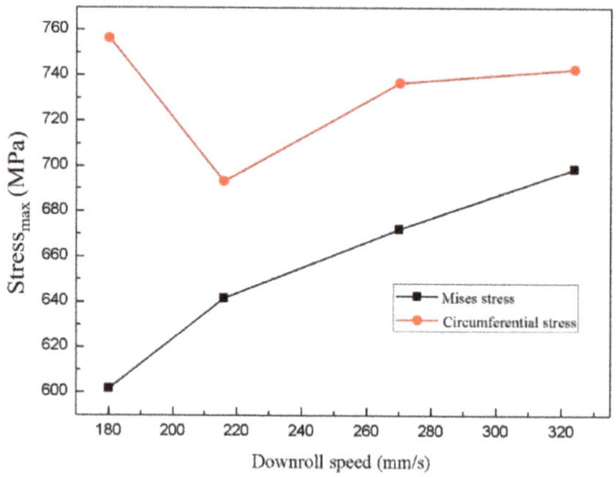

Figure 12. Effect of different rolling speeds on equivalent and circumferential stress.

4. Conclusions

This study uses finite element simulation to analyze the stress–strain distribution, thickness reduction, and microstructural changes during the rim roll-forming process of high-strength dual-phase steel DP590. The simulation results are consistent with the expectations, and the main findings are as follows:

1. Stress and Strain Concentration: Stress and strain are mainly concentrated in the welded region and heat-affected zones (especially the CGHAZ and FGHAZ), leading to local deformation and thickness reduction.

2. Thickness Reduction: Thickness reduction primarily occurs in the groove and flange sections. The reduction rate in the groove section is 3.83% after the first rolling step, 7.33% after the second step, and 17% after the third step, mainly due to material softening and carbon diffusion in the heat-affected zones.

3. Role of Microstructure: Microstructural characteristics significantly influence stress and strain distributions. The layered martensitic structure in the CGHAZ and FGHAZ causes high stress, while softening in the ICHAZ and SCHAZ leads to thickness reduction. Coarse grain and Widmanstatten structures in the welded area also contribute to stress concentration, affecting forming stability.

4. Parameter Selection: In the roll forming process, a higher friction coefficient can improve forming quality, and a friction coefficient of 0.4 is recommended. Additionally, the proper feed-speed and roll-speed ratio can enhance stability and stress control, with the optimized match by the feed speed of 15 mm/s and a 1:1.2 ratio of the upper and lower roll speeds.

This study provides new insights into the key regions and microstructural characteristics of high-strength dual-phase steel DP590 during rim roll forming. By optimizing process parameters and conducting in-depth investigations into microstructures, the manufacturing quality and stability of rims can be enhanced. Future research directions can focus on the following areas:

1. Expansion into application fields: The successful application of dual-phase steel in rim manufacturing opens up potential for its use in other mechanical structures. Particularly in the automotive and aerospace industries and in rail transportation sectors, the high strength and excellent formability of dual-phase steel can significantly improve the strength and durability of structural components.

2. Exploration of new simulation methods and material types: Future studies could introduce multi-physics coupling simulations and advanced numerical optimization methods to optimize parameter selection and process design in rim-forming processes. Similarly, similar analytical techniques could be applied to other advanced high-strength steel materials, such as improved models of dual-phase steel or other complex structural alloy steels.

3. Further improvement of material performance: Research efforts could focus on enhancing mechanical properties and forming quality of dual-phase steel through microstructure control and process parameter optimization. This includes optimizing forming processes to reduce stress concentration and thickness reduction, thereby improving product stability and reliability.

Author Contributions: J.S.: supervision, methodology, writing—review and editing; J.L.: data curation, formal analysis, writing—original draft; L.Z.: resources, methodology; Z.J.: resources, investigation; Z.Z.: investigation, supervision; J.H.: methodology, conceptualization, writing—review and editing; C.M.: methodology, investigation. All authors have read and agreed to the published version of the manuscript.

Funding: This research received no external funding.

Institutional Review Board Statement: Not applicable.

Informed Consent Statement: Not applicable.

Data Availability Statement: The original contributions presented in the study are included in the article, further inquiries can be directed to the corresponding authors.

Conflicts of Interest: Author Zhiqiang Zhang is employed by the company Han-Steel, and Author Cheng Ma is employed by the company HBIS Group. The remaining authors declare that the research was conducted in the absence of any commercial or financial relationships that could be construed as a potential conflict of interest.

References

1. Zhang, W.; Xu, J. Advanced Lightweight Materials for Automobiles: A Review. *Mater. Des.* **2022**, *221*, 110994. [CrossRef]
2. Gardie, E.; Paramasivam, V.; Dubale, H.; Tefera Chekol, E.; Selvaraj, S.K. Numerical Analysis of Reinforced Carbon Fiber Composite Material for Lightweight Automotive Wheel Application. *Mater. Today Proc.* **2021**, *46*, 7369–7374. [CrossRef]
3. Deepati, A.K.; Alhazmi, W.; Benjeer, I. Mechanical Characterization of AA5083 Aluminum Alloy Welded Using Resistance Spot Welding for the Lightweight Automobile Body Fabrication. *Mater. Today Proc.* **2021**, *45*, 5139–5148. [CrossRef]
4. Sharma, A.; Yadav, R.; Sharma, K. Optimization and Investigation of Automotive Wheel Rim for Efficient Performance of Vehicle. *Mater. Today Proc.* **2021**, *45*, 3601–3604. [CrossRef]
5. Li, Z.; Wu, D.; Lv, W.; Kang, S.; Zheng, Z. The Effects of Thermomechanical Processing on the Microstructure and Mechanical Properties of Ultra-High Strength Dual Phase Steel. *Adv. Mater. Res.* **2013**, *631–632*, 666–669. [CrossRef]
6. Pütz, F.; Fehlemann, N.; Göksu, V.; Henrich, M.; Könemann, M.; Münstermann, S. A Data Driven Computational Microstructure Analysis on the Influence of Martensite Banding on Damage in DP-Steels. *Comput. Mater. Sci.* **2023**, *218*. [CrossRef]
7. Azizi, H.; Samei, J.; Zurob, H.S.; Wilkinson, D.S.; Embury, D. A Novel Approach to Producing Architectured Ultra-High Strength Dual Phase Steels. *Mater. Sci. Eng. A* **2022**, *833*, 142582. [CrossRef]
8. Saha, D.C.; Biro, E.; Gerlich, A.P.; Zhou, Y. Influences of Blocky Retained Austenite on the Heat-Affected Zone Softening of Dual-Phase Steels. *Mater. Lett.* **2020**, *264*, 127368. [CrossRef]
9. González-Zapatero, W.F.; Rosado-Carrasco, J.G.; Ambriz, R.R.; Jaramillo, D. Low Cycle Fatigue Properties Assessment and Damage Influence on DP 500/800 Steel Sheet. *J. Mater. Res. Technol.* **2023**, *23*, 2231–2243. [CrossRef]
10. Krishnamraju, M.; Reddy, P.V.; Appalanaidu, B.; Markendeya, R. Mechanical Behavior and Forming Characteristics of Tailor-Welded Blanks of Structural Materials: A Review. *Multiscale Multidiscip. Model. Exp. Des.* **2024**. [CrossRef]
11. Ahmad, M.N.; Kamil, A.I.; Muhamad, W.M.W. Application of Shape Optimization Method on Steel Wheel Rim. *Appl. Mech. Mater.* **2014**, *564*, 13–18. [CrossRef]

12. Satheeshkumar, V.; Narayanan, R.G.; Gunasekera, J.S. Chapter 3—Sustainable Manufacturing: Material Forming and Joining. In *Sustainable Manufacturing Processes*; Narayanan, R.G., Gunasekera, J.S.B.T.-S.M.P., Eds.; Elsevier: Amsterdam, The Netherlands, 2023; pp. 53–112. ISBN 978-0-323-99990-8.
13. Kim, D.C.; So, W.J.; Kang, M.J. Effect of Flash Butt Welding Parameters on Weld Quality of Mooring Chain. *Arch. Mater. Sci. Eng.* **2009**, *38*, 112–117.
14. Çetinkaya, C.; Arabaci, U. Flash Butt Welding Application on 16MnCr5 Chain Steel and Investigations of Mechanical Properties. *Mater. Des.* **2006**, *27*, 1187–1195. [CrossRef]
15. Arabaci, U.; Çetînkaya, C.; Akay, A. An Investigation of Effects of Upsetting Current Time and Normalization Heat Treatment on Mechanical Properties of X40CrMoV5 1 and Ç1030 Steels Joined by Flash Butt Welding. *Mater. Des.* **2007**, *28*, 2351–2360. [CrossRef]
16. Yang, Z.; Zhou, W.L. The FEM Numerical Simulation of Die Spinning with Three-Dimensional. *Key Eng. Mater.* **2013**, *567*, 81–86. [CrossRef]
17. Gupta, A.; Pradhan, S.K.; Bajpai, L.; Jain, V. Numerical Analysis of Rubber Tire/Rail Contact Behavior in Road Cum Rail Vehicle under Different Inflation Pressure Values Using Finite Element Method. *Mater. Today Proc.* **2020**, *47*, 6628–6635. [CrossRef]
18. Kiuchi, M.; Kenji, A.; Onodera, R. Computerized Numerical Simulation. *CIRP Ann. Manuf. Technol.* **1995**, *44*, 239–242. [CrossRef]
19. Abe, Y.; Mori, K.; Ebihara, O. Optimisation of the Distribution of Wall Thickness in the Multistage Sheet Metal Forming of Wheel Disks. *J. Mater. Process. Technol.* **2002**, *125–126*, 792–797. [CrossRef]
20. Bi, D.; Yang, G.; Chu, L.; Zhang, J.; Wang, Z. Numerical Simulation on Spinning Forming Process of Automotive Wheel Rim. *Mater. Sci. Forum* **2012**, *704–705*, 1458–1464. [CrossRef]
21. Fang, G.; Gao, W.R.; Zhang, X.G. Finite Element Simulation and Experiment Verification of Rolling Forming for the Truck Wheel Rim. *Int. J. Precis. Eng. Manuf.* **2015**, *16*, 1509–1515. [CrossRef]
22. Lu, P.; Zhang, Y.K.; Ma, F. Finite Element Analysis on Multi-Step Rolling Process and Controlling Quality Defect for Steel Wheel Rim. *Adv. Mech. Eng.* **2015**, *7*. [CrossRef]
23. Hwang, J.K. Thermal Behavior of a Rod during Hot Shape Rolling and Its Comparison with a Plate during Flat Rolling. *Processes* **2020**, *8*, 327. [CrossRef]
24. *GB/T 31961-2015*; Rims for Truck-Bus. National Standard of the People's Republic of China: Beijing, China, 2015.
25. Naderi, G.; Dibajian, S.H.; Torshizi, S.E.M.; Masoumian, A.; Stockinger, M. An ABAQUS Plugin for Rotary Draw Bending of Tight Fit Pipes. *Int. J. Press. Vessel. Pip.* **2022**, *200*, 104827. [CrossRef]
26. Ya, S.; Eisenträger, S.; Qu, Y.; Zhang, J.; Kuen, T.; Song, C. Seismic Analysis of Post-Tensioned Concrete Gravity Dams Using Scaled Boundary Finite Elements Implemented as ABAQUS UEL. *Soil Dyn. Earthq. Eng.* **2023**, *164*, 107620. [CrossRef]
27. Sabik, A.; Witkowski, W. On Implementation of Fibrous Connective Tissues' Damage in Abaqus Software. *J. Biomech.* **2023**, *157*, 111736. [CrossRef]
28. Liu, P.R.; Lu, P.; Lv, P.J.; Lin, X.F. Numerical Simulation of Flaring of 590MPa High Strength Steel Rim. In Proceedings of the 15th Annual Meeting of the National Society of Plastic Engineering and the 7th Global Chinese Plastic Processing Technology Exchange Conference, Jinan, China, 13–16 October 2017; pp. 1037–1040.
29. Song, J.; Zhu, L.; Wang, J.; Lu, Y.; Ma, C.; Han, J.; Jiang, Z. Physical Simulation and Numerical Simulation of Flash Butt Welding for Innovative Dual Phase Steel DP590: A Comparative Study. *Materials* **2023**, *16*, 3513. [CrossRef]
30. Chen, H.; Wang, X.; Song, R.; Wang, Y.; Huo, W.; Zhang, Y.; Zhang, S. Microstructure, Mechanical Properties and Strengthening Mechanism of High Strength Hot-Rolled Ferrite-Martensite Dual Phase Steel. *Mater. Charact.* **2024**, *207*, 113545. [CrossRef]
31. Basu, S.; Patra, A.; Jaya, B.N.; Ganguly, S.; Dutta, M.; Samajdar, I. Study of Microstructure-Property Correlations in Dual Phase Steels for Achieving Enhanced Strength and Reduced Strain Partitioning. *Materialia* **2022**, *25*, 101522. [CrossRef]
32. Sodjit, S.; Uthaisangsuk, V. Microstructure Based Prediction of Strain Hardening Behavior of Dual Phase Steels. *Mater. Des.* **2012**, *41*, 370–379. [CrossRef]

materials

MDPI

rticle

Crashworthiness Performance and Multi-Objective Optimization of Bi-Directional Corrugated Tubes under Quasi-Static Axial Crushing

iuxiao Zou [1], Xin Wang [1], Ruojun Wang [1], Xin Huang [1], Menglei Li [1], Shuai Li [2], Zengyan Jiang [3,*] and Veilong Yin [1,4,*]

1 Center for Composite Materials and Structures, Harbin Institute of Technology, Harbin 150080, China; zlx19851196807@163.com (L.Z.)
2 Department of Engineering Mechanics, Kunming University of Science and Technology, Kunming 650031, China
3 Aerodynamics Research Institute, AVIC (Aviation Industry Corporation of China), Harbin 150080, China
4 National Key Laboratory of Science and Technology on Advanced Composites in Special Environments, Harbin Institute of Technology, Harbin 150080, China
* Correspondence: jiangzyan@avic.com (Z.J.); ywl@hit.edu.cn (W.Y.)

Abstract: Longitudinal corrugated tubes (LCTs) exhibit stable platform force under axial compression but have low specific energy absorption. Conversely, circumferential corrugated tubes (CCTs) offer higher specific energy absorption but with unstable platform force. To overcome these limitations, this paper introduces a novel bi-directional corrugated tube (BCT) that amalgamates the strengths of both the CCT and LCT while mitigating their weaknesses. The BCT is formed by rolling a bi-directional corrugated structure into a circular tubular form. Numerical simulations of the BCT closely align with experimental results. The study further examines the influence of discrete parameters on the BCT's performance through simulations and identifies the tube's optimal design using the integral entropy TOPSIS method. A full factorial experimental approach is then employed to investigate the impact of radial amplitude, axial amplitude, and neutral surface diameter on the crushing behavior of the BCT, comparing it with the CCT and LCT. The results reveal that increasing A_i enhances the axial resistance of the structure, while increasing A_j reduces the buckling effect, resulting in a higher specific energy absorption and lower ultimate load capacity for the BCT compared to the CCT and LCT. A simultaneous multi-objective optimization of the CCT, LCT, and BCT confirms that the BCT offers superior specific energy absorption and ultimate load capacity. The optimal configuration parameters for the BCT have been determined, providing significant insights for practical applications in crashworthiness engineering.

Keywords: bi-directional corrugated tubes; axial crushing; energy absorption; stable plateau force; multi-objective optimization

1. Introduction

In the quest for light weight and high energy absorption structures, thin-walled tubes have emerged as the predominant choice for energy absorption applications [1–3], notably in spacecraft landers [4], medical apparatus [5], robots [6], and automobiles [7–9]. An exemplary energy-absorbing structure must possess not only a superior capacity to absorb energy but also maintain a consistent and stable reaction force throughout its deformation process. This combination of attributes ensures effective energy dissipation while minimizing the impact on the system or its occupants.

Thin-walled tubes subjected to axial compression are frequently utilized as crash boxes behind car bumpers, owing to their extended deformation stroke and superior energy absorption. The performance of these tubes is predominantly influenced by the topological characteristics of their cross-sections and their subsequent evolution, which can be meticulously controlled through mathematical functions or curves. This approach allows for the facile generation of diverse cross-sectional profiles by adjusting specific parameters of the control function or curve. While Wu et al. [10] introduced the use of the Fourier function in cross-section design, it was noted that Fourier cross-section tubes may have limited applicability in sandwich structures due to non-coincidence of the tangent points of their concave sections along the same circumference. Alternatively, Deng et al. [11,12] proposed circumferentially corrugated tubes, which exhibit enhanced energy absorption capabilities due to the increased structural resistance conferred by their complex cross-sections. Similarly, Li et al. [13] and Eyvazian et al. [14] introduced transversely corrugated square tubes with sinusoidal profiles, demonstrating higher specific energy absorption compared to conventional square tubes. However, it is noteworthy that these circumferential corrugated tubes, despite their high energy absorption, often produce force–displacement curves characterized by high initial peak loads and unstable plateau forces. This instability stems from the repeated buckling and bending failures of axially loaded tubes.

Extensive exploratory studies have been conducted to mitigate the initial peak force and enhance the stability of the crushing process in thin-walled structures. Singace et al. [15] experimentally investigated the energy absorption characteristics of corrugated tubes, introducing corrugations to enforce plastic deformation and improve the uniformity of the load–displacement behavior under axial loading. Eyvazian et al. [16] explored the impact of corrugation on the crushing behavior of circular aluminum tubes, focusing on energy absorption and damage patterns. Their findings indicated that longitudinal corrugated tubes promote more predictable and controllable collapse patterns. The controllability and predictability of the collapse patterns are crucial for applications where precise energy management is essential, such as in automotive safety systems, aerospace structures, and protective packaging. By utilizing corrugated tubes, engineers can design systems that more effectively dissipate impact energy, reducing the risk of injury or damage to occupants or surrounding structures. Consequently, these tubes are considered promising candidates for controlled energy absorption devices and have garnered significant research interest.

Dong et al. [17] designed flexure-inducing tubes based on the modal analysis of straight-walled triangular tubes, enabling folding according to specific modes by strategically presetting convex and concave structural forms. Chen et al. [18] analyzed the non-linear elastic–plastic behavior of circular tubes with corrugated profiles using finite element methods, providing insights into their deformation mechanisms and energy absorption characteristics. Eyvazian et al. [19] conducted both experimental and theoretical examinations of corrugated tubes under axial loading, validating their findings and reinforcing the understanding of their crushing behavior. Alkhatib et al. [20] focused on the additive manufacturing and experimental testing of sinusoidal corrugated tubes under quasi-static loading conditions. Imai et al. [21] employed isogeometric analysis and experiments to study the effect of contour curvature on the energy absorption characteristics of cylindrical corrugated tubes. Wu et al. [22] proposed sinusoidal corrugated tubes designed to control collapse modes and minimize the initial peak crushing force and load fluctuation. Their findings revealed that while these tubes exhibited more uniform force–displacement curves, they significantly reduced energy absorption compared to conventional straight circular tubes. Inspired by the concept of functionally graded materials, Yao et al. [23] developed novel hybrid corrugated tubes with an amplitude profile that decreases progressively from large to small during compression. This design was found to enhance energy absorption capacity compared to ordinary corrugated tubes. Finally, Ha et al. [24] introduced an innovative tubular corrugated structure, modeled after the profile of a coconut tree. This design aimed to improve energy absorption, reduce the initial peak crushing force, and stabilize the crushing process.

Despite their remarkable ability to effectively mitigate initial peak forces and plateau force fluctuations, corrugated tubes exhibit a trade-off in the form of reduced specific energy absorption. The results indicate that deformation behavior can be stabilized and load–displacement curves, as well as energy absorption, can be controlled by introducing corrugations with varying distances. In summary, these corrugated tubes demonstrate significant potential for reducing initial peak forces and enhancing deformation stability, which has garnered increasing attention. However, it is important to note that the specific energy absorption of these corrugated tubes is diminished.

As illustrated in Figure 1, longitudinally corrugated tubes (LCTs) exhibit a low collapsing force, smooth plateau force, and reduced energy absorption under axial compression. In contrast, circumferentially corrugated tubes (CCTs) demonstrate higher energy absorption, accompanied by a higher collapsing force, and a fluctuating plateau force. Based on the shape of the force–displacement curve under quasi-static compression, energy absorbers can be broadly classified into two categories: those with a nearly "flat-topped" load–displacement curve and those with an initial peak force followed by a steeply decreasing curve. Obviously, the force–displacement curve of the LCT aligns with the former type, characterized by a nearly "flat-topped" profile, whereas the CCT exhibits the latter type, marked by an initial peak force followed by a steep decline. This dichotomy underscores the contrasting force–displacement characteristics of the LCTs and CCTs. An analysis of the cross-sectional shapes reveals that the ripples in the LCTs are present on the sidewall, while those in the CCTs appear on the cross-section. By integrating sidewall corrugations and cross-section corrugations, a bi-directional corrugated tube (BCT) can be developed. Notably, BCTs exhibit a unique tunability, as their properties can be modulated by adjusting the amplitudes of the cross-sectional and sidewall corrugations. When the cross-sectional corrugation amplitude of the BCT is zero, it degenerates into an LCT; similarly, when the sidewall corrugation amplitude of the BCT is 0, it degenerates into a CCT. This adaptability suggests the feasible of identifying BCT structures with relatively stable plateau forces and higher energy absorption.

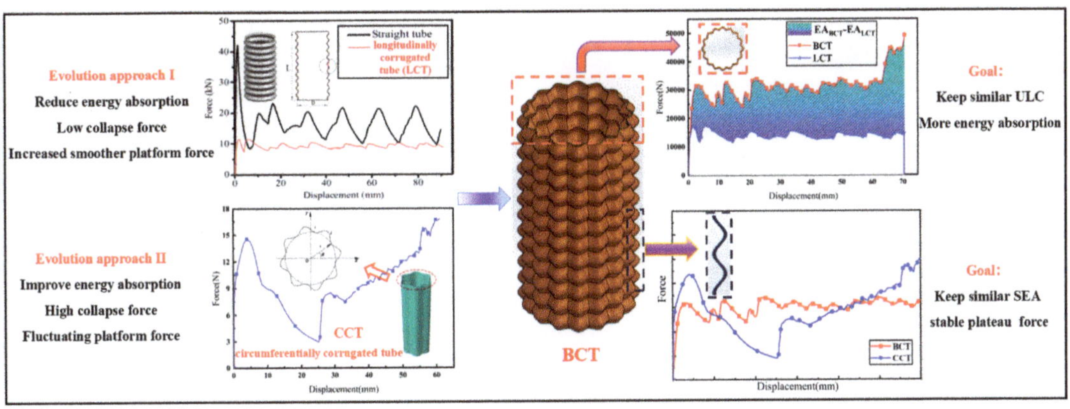

Figure 1. Comparing the force–displacement curves of a CCT and an LCT under axial compressive loads, and in order to keep similar specific energy absorption and obtain a more stable plateau force, a BCT structure is proposed.

In this study, BCTs were fabricated using 3D printing and subjected to axial compression tests to investigate their deformation mechanisms. To further explore the deformation and energy-absorbing properties of BCTs under axial compression, a comprehensive finite element model was developed using the commercial software ABAQUS. Initially, discrete parameters were parameterized based on the finite element method to assess the crashworthiness of these BCTs. The integral entropy TOPSIS method was then employed to evaluate and select the optimal discrete parameters for crashworthiness. Subsequently, the full factorial design method was utilized to examine the effects of continuous parameters such as circumferential ripple amplitude A_i, axial ripple amplitude A_j, and neutral plane diameter D. The results were compared with those of CCTs and LCTs to underscore the advantages of BCTs. Finally, multi-objective optimization was performed for CCTs, LCTs, and BCTs to highlight the combination of high specific energy absorption and stable plateau force in BCTs, ultimately determining the optimal geometric parameters for the tubes.

2. Materials and Methods

2.1. Geometric Features

Figure 2a meticulously illustrates the bi-directional corrugated structure, showcasing its unique undulating patterns in two perpendicular directions. Within this context, the Cartesian coordinate system, defined by the axes (x, y, z), serves as a fundamental reference framework for elucidating the geometry of the structure. The governing equations in the x and z directions are:

$$y_i = A_i \times cos[2\pi Nx/(\pi D)] \quad (1)$$

$$y_j = A_j \times cos[2\pi Mz/H] \quad (2)$$

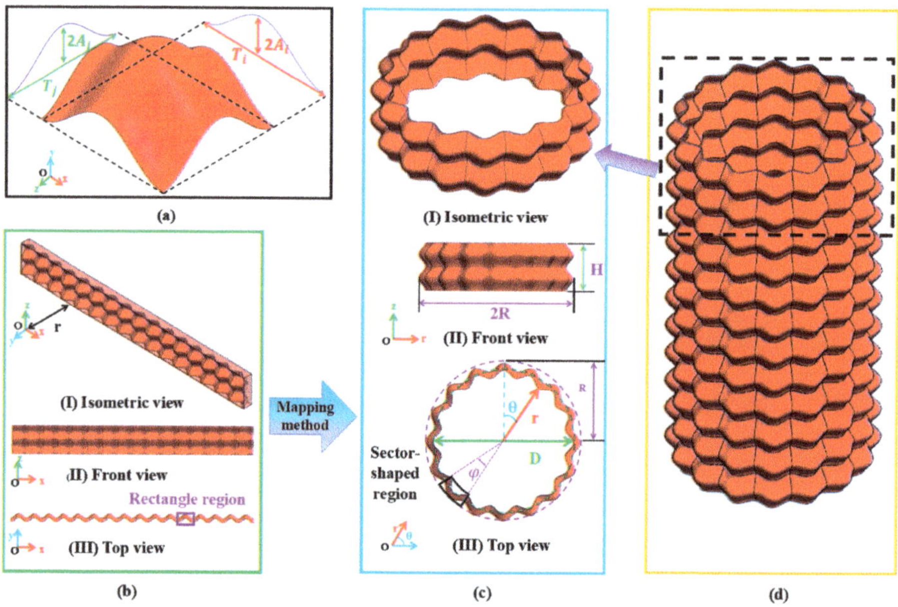

Figure 2. Geometry modeling process for the BCT: (**a**) bi-directional corrugated structure, (**b**) structure with 12 × 2 cells, (**c**) BCT modeling by the mapping method, (**d**) geometric modeling of the BCT.

The amplitudes of the ripple curves in the x-axis and z-axis directions are denoted as A_i and A_j, respectively. N represents the number of units in a single-layer BCT, while D denotes the diameter of the neutral plane of the BCT. M signifies the number of layers in the BCT, and H indicates the height of the BCT.

To fully leverage the superior mechanical properties of bi-directional corrugated structures within tubular frameworks, the bi-directional corrugated structure is integrated as the unit element within the tubular structure to enhance its resistance to axial compression. The fabrication process of the BCT is illustrated in Figure 2. It is evident that the 3D structure depicted in Figure 2b (I) is transformed into a circular tubular form as shown in Figure 2c (II). The complete BCT consists of N and M bi-directional corrugated units arranged in the circumferential and axial directions, respectively. Here, N and M denote the number of bi-directional corrugated structural units aligned in the x and z directions within the

Cartesian coordinate system. Figure 2b presents the isometric, front, and top views of this structure, with N and M values of 16 and 2, respectively. The distance from the XOZ plane is denoted as r, representing the inner diameter of the BCT, as illustrated in Figure 2b (I). The bi-directional corrugated structure in Figure 2b is bent to form the circular tubular BCT depicted in Figure 2c. The coordinates of the BCT in the column coordinate system are represented as (r, θ). The radial thickness of the BCT, shown in Figure 2c (II), is determined by $2(A_i + A_j)$, which also represents the thickness of the bi-directional corrugated structural unit along the x-axis. Furthermore, the bending angle φ is calculated as $\varphi = 2\pi/N$. The geometric model of a twelve-layer BCT is illustrated in Figure 2d, with specific geometric parameters detailed in Figure 2c. For clarity, the wall thickness (t) is 1 mm, and the length (H) is 120 mm.

2.2. Crashworthiness Indicators

This section examines six key crashworthiness criteria to offer a comprehensive understanding of energy absorbers through various metrics. These metrics are defined as follows.

1. Energy absorption (EA): This is defined as the area under the force–displacement curve, given by:

$$EA = \int_0^s F(x)dx \tag{3}$$

where $F(s)$ denotes the loading force as a function of displacements. S represents the effective stroke, which is numerically defined as 70% of the tube length.

2. Specific energy absorption (SEA): This metric represents the energy absorption per unit mass and indicates the energy absorption efficiency of the structure. It can be expressed as:

$$SEA = \frac{EA}{m} = \frac{\int_0^S F(x)dx}{m} \tag{4}$$

3. Initial peak crushing force ($IPCF$): This refers to the maximum force attained on the force–displacement curve following the elastic phase. The force at this peak is termed the $IPCF$.
4. Mean crushing force (MCF): MCF is defined as the ratio of EA to the effective stroke S. It can be quantified by the following equation:

$$MCF = \frac{EA}{S} = \frac{\int_0^S F(x)dx}{S} \tag{5}$$

5. Crushing force efficiency (CFE): ratio of MCF to F_{max}.

$$CFE = \frac{MCF}{F_{max}} \tag{6}$$

6. Undulation of load carrying capacity (ULC): ULC quantifies the smoothness of the force–displacement curve by calculating the ratio of the work done due to deviations of the actual load from the MCF to the EA. A smaller ULC value signifies a more uniform and smoother energy absorption process.

$$ULC = \frac{\int_0^S |F(x) - MCF|dx}{\int_0^S F(x)dx} \tag{7}$$

2.3. Compression Test

The material used for the BCT was stainless steel 316 L, with the following mechanical properties: [25] density $\rho = 7980$ kg/m^3, Young's modulus $E = 171$ GPa, yield stress $\sigma_y = 552$ MPa, and Poisson's ratio $\nu = 0.3$. Given that quasi-static loading conditions are simulated, strain rate effects are not considered into the finite element modeling. To evaluate the stability of the BCT platform under axial compression and its capacity for energy absorption, the BCT specimen was prepared as illustrated in Figure 3a. The specimen, fabricated using additive manufacturing techniques, features a height of 120 mm and a wall thickness of 1 mm. The chosen parameters for the number of corrugations N and the number of layers M were 16 and 12, respectively. Additionally, both A_i and A_j were set to 2 mm, and the diameter of the neutral plane (D) was specified as 51 mm.

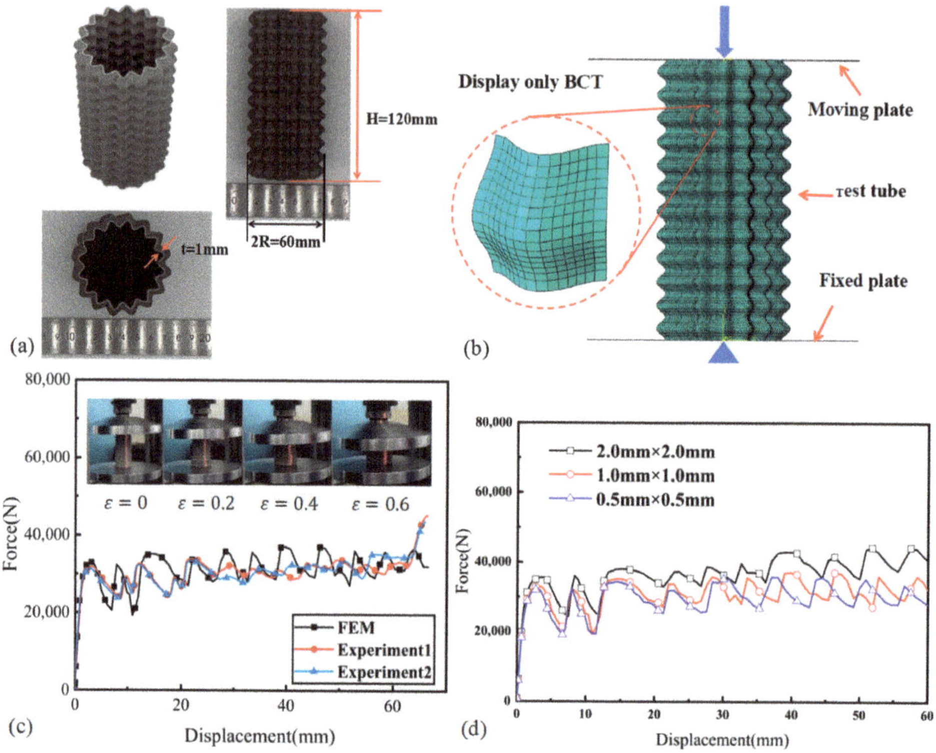

Figure 3. (**a**) Schematic diagram of the BCT specimen. (**b**) FE model of the tube under axial impact loading. (**c**) Force–displacement curves of experimental and numerical results of the BCT. (**d**) Mesh-size validation for the force–displacement of the BCT under axial compression.

The degradation history of the BCT, as depicted in Figure 3c, provides valuable insights into its behavior under axial compression loads applied at a quasi-static rate of 5 mm/min. As compression displacement increases, the BCT undergoes a controlled deformation process, with layers progressively collapsing from the base upwards. This compression includes the formation of lobes, with 12 folds present and observable. The force–displacement curves for the BCT, as shown in Figure 3c, demonstrate a high degree of correlation between experimental and simulated data. These curves reveal a stable plateau force throughout the compression process. Notably, the experimental force–displacement curves exhibit reduced fluctuation, with the number of oscillations corresponding to the number of pre-folded layers. This finding further supports the notion that the BCT's deformation is well ordered and controlled, leading to a more stable and predictable performance under compression.

2.4. FE Modeling

The crashworthiness of the BCT was assessed using the finite element analysis software ABAQUS/Explicit. The finite element model, as depicted in Figure 3b, consists of three components: the moving plate, the fixed plate, and the test tube. Initially, all degrees of freedom of the rigid plate are coupled to a reference point, where boundary conditions are applied. The upper rigid plate is subjected to a vertical downward displacement with a velocity of 1 m/s [22], while the lower rigid plate is fixed in all six degrees of freedom. Meshing is performed using 4-node simplified integral linear quadrilateral elements (S4R). The numerical model incorporates generalized contact with a friction coefficient of 0.3 to accurately represent the interaction between the rigid plate and the tube. To balance computational efficiency with accuracy, a cell size of 1.0 mm × 1.0 mm is utilized. For the rigid plates, which are modeled as discrete rigid shells, no cell properties are assigned, and the mesh size for these shells is set to 4 mm. This approach allows for a detailed analysis of the BCT's deformation and energy absorption mechanisms while keeping the overall simulation time manageable.

As shown in Figure 3c, the force–displacement curves obtained from the FEM exhibit an elastic phase similar to the experiment. After reaching a similar $IPCF$, the curve fluctuates up and down in the plastic plateau segment in small magnitudes. The overall agreement between the FEM and experimental force–displacement curves, particularly in the elastic and plastic plateau segments, provides confidence in the validity and accuracy of the simulation results. These results can then be used to further analyze the crashworthiness of the BCT material.

Figure 3c illustrates a comparison between the experimental force–displacement data and numerical results, demonstrating a strong concordance between the finite element simulations and the experimental observations. The response curves plotted in Figure 3c exhibit comparable trends and fluctuation characteristics. The $IPCF$ for specimen 1 and specimen 2 are 31.76 kN and 31.26 kN, respectively, while the simulated $IPCF$ is 32.95 kN, as detailed in Table 1. This close correspondence further reinforces the validity of the finite element model. Moreover, the discrepancies in EA and MCF between experimental and simulated data are within 3%, highlighting the precision of the simulation in capturing the essential characteristics of the BCT. These findings provide strong evidence that the finite element model was successfully calibrated to accurately represent the mechanical behavior of the BCT under the given loading conditions.

Table 1. Error analysis of experiment and simulation.

	Mass (g)	*MCF* (kN)	*EA* (J)	*SEA* (J/g)
specimen 1	243.3	30.37	1974.13	8.11
specimen 2	242.8	30.42	1977.18	8.14
FEM	240.99	31.00	2014.98	8.36

3. Discussion

3.1. Discrete Parametric Analysis

3.1.1. Effect of N and M

Figure 4 provides a detailed analysis of 25 design cases with varying values for N and M, revealing that the deformation modes of the BCTs predominantly fall into three distinct categories: progressive flexure mode (P-mode) [26], diamond mode (D-mode), and mixed mode (M-mode). When $N = 20$, the collapse consistently occurs in P-mode, regardless of the value of M. This observation suggests that a higher N value promotes a more uniform progressive flexure due to increased rigidity and stability provided by the higher number of cells along the tube's length. For $N = 4$, both $M4$ and $M6$ exhibit D-mode deformation, similar to conventional corrugated tubes (CTs). The fewer layers in these configurations result in larger individual cell heights, making the structure more prone to diamond-shaped deformations under compression. As the number of layers increases (e.g., $M8$), the deformation transitions to P-mode. This indicates that a higher layer count, resulting in smaller cell heights, enhances the structure's ability to undergo progressive flexure rather than localized diamond-shaped buckling. Interestingly, in configurations with a greater number of layers (e.g., *M*10 and *M*12), a mixed deformation mode is observed. Initially, the structure undergoes P-mode deformation, followed by a transition to D-mode. This behavior can be attributed to the balance between the increased structural integrity provided by the additional layers and the inherent tendency of the structure to buckle in a diamond pattern under certain conditions. For intermediate values ($8 \leq N \leq 16$), the deformation mode shifts from mixed mode to P-mode as N increases. This shift indicates a critical threshold, where the structural configuration favors progressive flexure over mixed deformation, likely due to the optimized balance between cell height and layer count. These observations highlight the complex interplay between the geometric parameters N and M in determining the deformation behavior of BCTs. By leveraging this understanding, engineers and designers can tailor the geometric configuration of BCTs to achieve desired mechanical properties for specific applications.

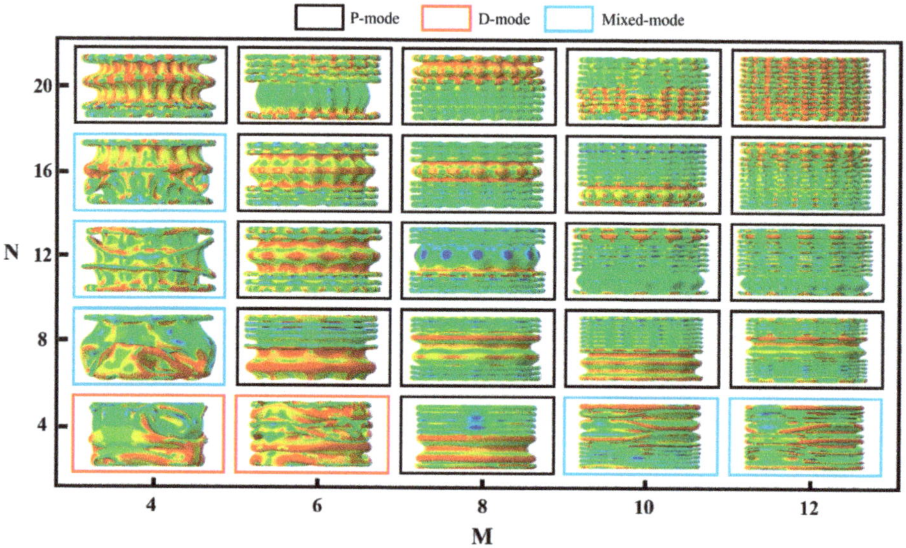

Figure 4. Distribution of deformation modes.

Figure 5 presents the load–displacement curves for 25 different cases, providing insight into the mechanical performance of BCTs under varying geometric parameters. Notably, the initial stiffness of the BCTs is observed to increase with a larger N. This phenomenon can be attributed to the greater number of cells along the tube's length, which enhances the overall structural rigidity and resistance to initial deformation. Conversely, the initial peak crushing force ($IPCF$) decreases with an increase in M or a decrease in N. This indicates that BCTs with more layers (higher M) or fewer cells along the length (lower N) are less capable of sustaining high peak forces, likely due to the reduced height of individual cells, which leads to earlier onset of buckling and less force being required for deformation. Except for cases *N*4*M*4 and *N*8*M*4, the peaks of the BCT curves are closely aligned, reflecting effective energy absorption capabilities. These two exceptions can be explained by their unique configurations where the combination of N and M results in less optimal energy dissipation characteristics, possibly due to localized buckling or irregular deformation patterns. Furthermore, as N decreases, the load–displacement curves exhibit reduced fluctuation, leading to smoother energy absorption. This behavior suggests that BCTs with fewer cells along the length distribute the applied load more evenly across the structure, thereby avoiding abrupt changes in force and resulting in a more stable energy absorption process. Specifically, for lower N values (e.g., $N4$), the curves show more pronounced peaks and valleys, indicating intermittent energy absorption due to localized buckling events. As N increases (e.g., $N12$ and above), the curves become smoother, with less pronounced fluctuations, suggesting a more uniform deformation process and continuous energy absorption. These findings underscore the importance of optimizing both N and M to achieve desired mechanical properties, particularly in applications requiring high energy absorption and minimal force fluctuations.

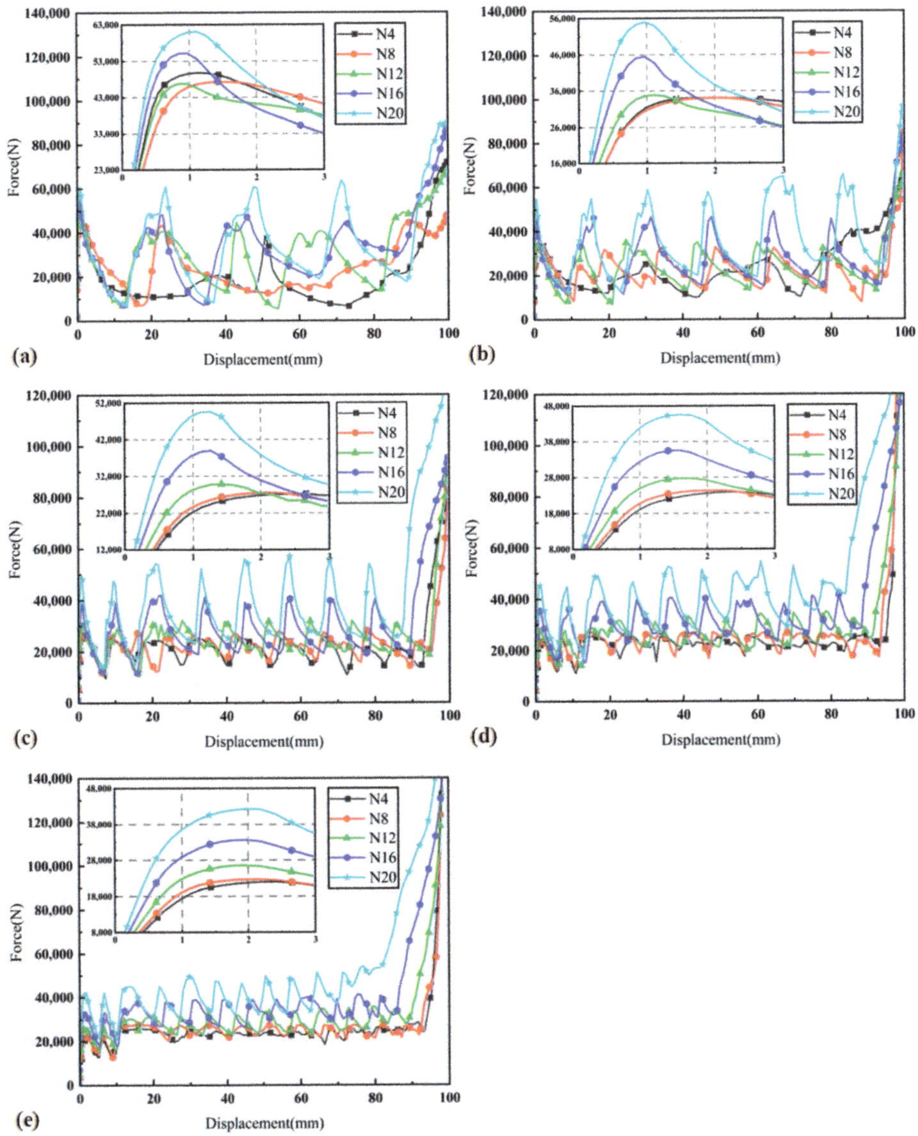

Figure 5. Load–displacement diagram of BCTs: (**a**) $M4$, (**b**) $M6$, (**c**) $M8$, (**d**) $M10$, and (**e**) $M12$.

Figure 6 presents a detailed analysis of the crashworthiness parameters of the BCTs, offering insights into their performance under various geometric configurations. *IPCF* increases with a decrease in *M* or an increase in *N*. This indicates that BCTs with fewer layers (lower *M*) or more cells along the length (higher *N*) can sustain higher peak forces before deformation. This behavior is likely due to the increased structural integrity provided by a higher number of cells, which enhances resistance to initial impact forces. *SEA* rises with increasing values of both *M* and *N*, though the influence of *N* is more pronounced. This suggests that while both parameters contribute to the energy absorption capacity of the BCTs, the number of cells along the length (*N*) plays a more critical role. By fostering a more distributed pattern of energy absorption, a higher *N* value leads to more distributed energy absorption, enhancing the overall crashworthiness of the structure. *ULC* shows minimal sensitivity to changes in *N* but increases significantly with higher *M*. This indicates that the layer count (*M*) is a more crucial factor in determining the maximum load-bearing capacity of the BCT before failure. Higher *M* values enhance the overall structural strength, allowing the BCT to absorb more energy before reaching its ultimate load. *CFE* generally increases with larger values of *M*. This reflects the improved ability of the BCT to absorb energy efficiently during deformation. Higher *M* values result in a more stable and controlled collapse mechanism, enhancing the overall energy absorption efficiency of the structure.

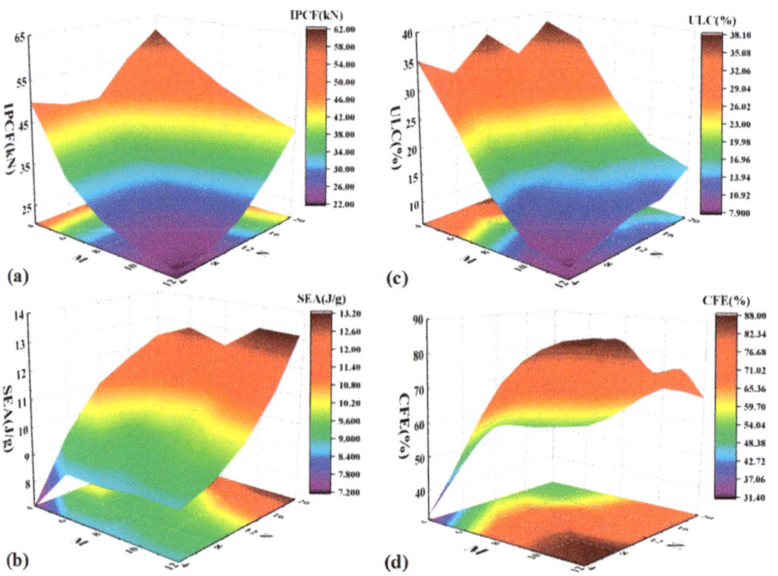

Figure 6. Crashworthiness parameter chart of the BCT: (**a**) *IPCF*, (**b**) *SEA*, (**c**) *ULC*, and (**d**) *CFE*.

The impact resistance of the BCT is also illustrated in Figure 6. The results indicate that BCTs with higher values of *N* and *M* exhibit superior impact resistance, demonstrating their effectiveness in dissipating impact energy and minimizing damage. A higher *N* increases *IPCF* due to the added structural elements, while a lower *M* decreases *IPCF* because of fewer layers to distribute the load.

Both *M* and *N* contribute positively to increasing *SEA*, but the impact of *N* is more significant, indicating a strong dependence on the longitudinal configuration of the BCT. On the other hand, the insensitivity of *ULC* to *N* suggests that the overall lengthwise configuration is less critical, whereas higher *M* enhances the BCT's ultimate load capacity significantly. Furthermore, larger *M* values result in higher *CFE*, showcasing better energy absorption capabilities and efficient crash management.

3.1.2. Selection Method and Results

Selecting the BCT from the 25 proposed designs presented a challenge due to the multiple evaluation criteria involved. To address this challenge, the Technique for Order of Preference by Similarity to Ideal Solution (TOPSIS) method [27], widely recognized for its efficacy and versatility across various domains, was employed. This method effectively mitigates subjective bias and accommodates both quantitative and qualitative data [27]. Utilizing the integral entropy TOPSIS approach [28], it is possible to derive the weights of the indicators w_i and calculate the relative proximity Q_j. Detailed procedural steps are provided in Appendix A.

Following the exclusion of six BCTs with ULC exceeding 30%, the remaining 19 BCTs were assessed based on crashworthiness metrics. Among these, SEA and CFE were identified as advantageous criteria, with weights assigned as 0.71 and 0.29, respectively. The relative distances between the tubes were calculated and are detailed in Appendix B (Table A1). Notably, $M12$ consistently yielded the highest Q_j irrespective of variations in N. When varying M, $N20$ and $N16$ consistently exhibited the two highest Q_j values. However, ULC for $N20$ was disproportionately high compared to $N16$. Consequently, for all subsequent analyses, $N16M12$ was selected as the optimal configuration.

3.2. Parametric Analysis of Continuous Type Parameters

3.2.1. Deformation Patterns

Table 2 outlines the design framework involving three variables, each evaluated across six levels, culminating in a total of 216 design cases. These cases incorporate various values for A_i, A_j, and D, which are subsequently simulated. An interesting aspect of this design framework is the ability to transform the BCT into other common structural forms by setting specific values for A_i and A_j. For instance, when both A_i and A_j are set to 0, the BCT effectively transforms into a circular tube (CT). This transformation allows for a direct comparison between the performance of the BCT and that of a traditional circular tube. Likewise, setting A_i to 0 results in the BCT resembling a LCT, while setting A_j to 0 yields a CCT.

Table 2. Values of each level of the parameters.

Parameters	Levels	Values
A_i	6	0, 0.5, 1.0, 1.5, 2.0, 2.5
A_j	6	0, 0.5, 1.0, 1.5, 2.0, 2.5
D	6	48, 52, 56, 60, 64, 68

Figure 7 presents a detailed depiction of the conformational distribution of BCTs ($D60$). Figure 7a displays the configuration of BCTs with 1 CT illustrated in black wireframe, 5 LCTs in green wireframe, 5 CCTs in blue wireframe, and 25 BCTs in red wireframe. Figure 7b elucidates the distribution of deformation modes observed in the $D60$ configuration. When A_i is 0, LCTs exhibit three distinct deformation modes. Similarly, when A_j is 0, CCTs exhibit two deformation modes. This analysis confirms that BCTs are capable of exhibiting a total of five distinct deformation modes.

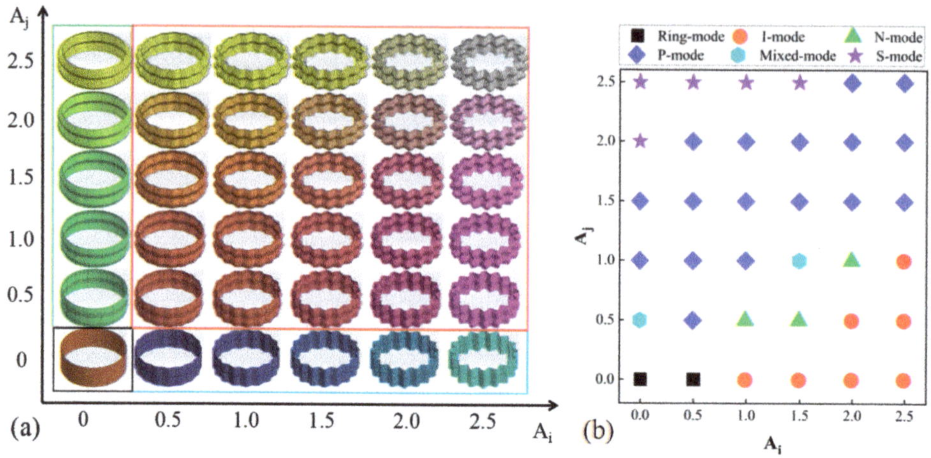

Figure 7. (**a**) BCT conformational distribution (*D*60) and (**b**) distribution of deformation patterns (*D*60).

Figure 8 illustrates the deformation modes and corresponding force–displacement curves for the BCT with *D*60, highlighting distinct behaviors under different configurations. Observed in $A_i 2.5 A_j 0.5$, this mode is characterized by high *IPCF* and unstable energy absorption within the plastic deformation region. The force–displacement curves show significant fluctuations, indicating inefficient energy absorption. This behavior is consistent with previous reports [29], where the structural instability leads to rapid changes in force. When $A_i 2.0 A_j 1.0$, the BCT exhibits *N*-mode, resulting in a reduction in *IPCF*. However, the force–displacement curves still display pronounced fluctuations. This indicates that while the peak force is lower, the energy absorption process remains unstable, leading to less efficient performance [18]. In the configuration $A_i 1.0 A_j 1.5$, the BCT exhibits P-mode, characterized by a lower *IPCF* and reduced fluctuations in the force–displacement curves. The plastic section shows a plateau force marginally exceeding *IPCF*, indicating a more stable and efficient energy absorption mode. This mode ensures that the energy is absorbed progressively, minimizing the risk of sudden structural failure. Observed in $A_i 1.5 A_j 1.0$, this mode demonstrates localized diamond deformation following P-mode. The force–displacement curves are similar to N-mode but with reduced fluctuation, indicating a combination of progressive and localized deformation mechanisms. This mixed behavior provides a balance between stability and energy absorption efficiency. When the corrugation parameters are set to $A_i 0.5 A_j 2.5$, the BCT exhibits S-mode, characterized by concurrent deformation of all units. This results in very smooth force–displacement curves with an ultimate load capacity (*ULC*) below 10%. The uniform deformation leads to highly efficient energy absorption with minimal fluctuations, making this mode the most stable and effective among the observed configurations. Overall, the analysis shows that by carefully selecting the corrugation parameters A_i and A_j, engineers can fine-tune the deformation modes and energy absorption capabilities of the BCT to suit specific design requirements.

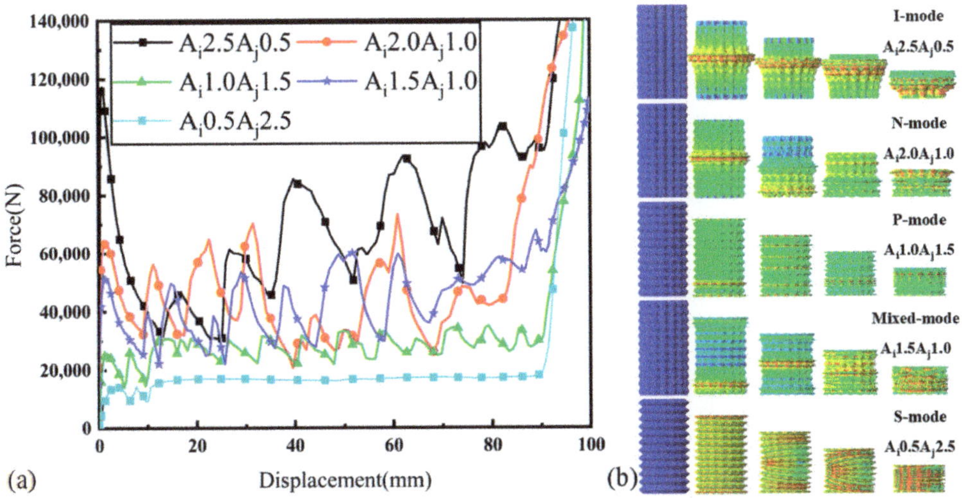

Figure 8. (**a**) Representative force–displacement curves and (**b**) deformation history diagram of representative deformation patterns.

3.2.2. Crashworthiness Analysis

Figure 9 presents the contour plots of specific energy absorption (SEA) for BCTs with varying diameters, providing a detailed visualization of how SEA varies with different geometric parameters A_i and A_j. The contour plots indicate that the lowest SEA values are observed at lower A_i and higher A_j values. This suggests that configurations with smaller initial cell heights (A_i) and larger lateral expansions (A_j) are less effective at absorbing energy, likely due to less optimal structural configurations for energy dissipation. As A_i increases or A_j decreases, SEA progressively increases. This indicates that increasing the initial cell height or reducing the lateral expansion enhances the energy absorption capability of the BCTs. Higher A_i values likely contribute to greater structural integrity and improved energy distribution during deformation, while lower A_j values reduce the spread of deformation, concentrating energy absorption in a more controlled manner. The intensification of the red regions on the contour plots as the diameter of the BCTs increases highlights an important aspect of their energy absorption performance. The deeper red hues signify higher SEA values, indicating that larger diameter BCTs exhibit greater energy absorption capabilities. This trend can be attributed to the increased material volume and structural capacity of larger diameter tubes, which provide more room for deformation and dissipation of impact forces. Additionally, the red regions shift towards higher values of A_i with increasing BCT diameter. This trend indicates that for larger diameter BCTs, higher initial cell heights (A_i) become more critical in achieving maximum SEA. This shift reflects the necessity for stronger structural configurations in larger diameter tubes to manage and distribute the absorbed energy efficiently.

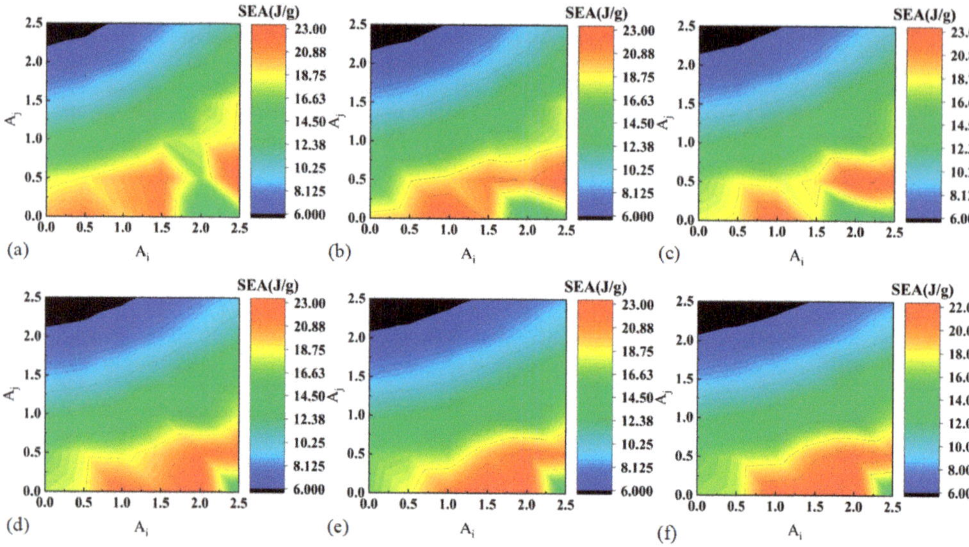

Figure 9. Contour plots of SEA : (**a**) $D48$, (**b**) $D52$, (**c**) $D56$, (**d**) $D60$, (**e**) $D64$, and (**f**) $D68$.

Figure 10 presents the contour plots of ultimate load capacity (ULC) for BCTs with varying diameters, providing detailed insights into how ULC is influenced by the geometric parameters A_i and A_j. The analysis reveals that ULC decreases with increasing A_j and decreasing A_i. This indicates that larger amplitude of axial ripples (A_j) and smaller amplitude of transverse ripple (A_i) lead to lower ULC, suggesting reduced structural strength under compressive loads. Notably, A_j has a more pronounced effect on ULC compared to A_i, highlighting its significant role in determining the compressive properties of the BCTs. The contour plots show that the regions of lowest ULC are concentrated within the ranges of $1.5 \leq A_j \leq 2.5$ and $0 \leq A_i \leq 1.5$. These ranges indicate the geometric configurations that result in the weakest structural performance under compressive loads. Understanding these ranges is crucial for avoiding designs that may lead to premature failure or inadequate load-bearing capacity. As the diameter of the BCT increases, the area corresponding to the minimum ULC values expands towards higher A_i values. This trend suggests that larger diameters require higher A_i to maintain structural integrity and avoid significant reductions in ULC. This shift highlights the need to adjust geometric parameters proportionally with the diameter to ensure optimal performance. A_j plays a pivotal role in modulating the compressive properties of the BCTs. Its significant impact on reducing ULC compared to A_i underscores the importance of lateral expansion in determining the overall structural strength. For optimal deformation stability, A_j should be selected at a higher value to minimize buckling effects. This ensures that the BCT can sustain higher compressive loads without experiencing instability or failure.

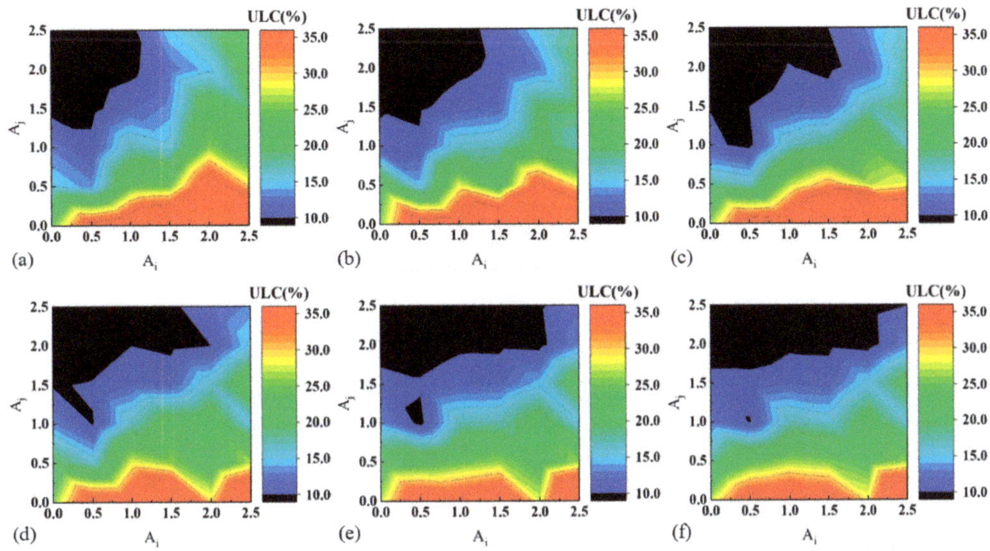

Figure 10. Contour plots of *ULC*: (**a**) *D*48, (**b**) *D*52, (**c**) *D*56, (**d**) *D*60, (**e**) *D*64, and (**f**) *D*68.

3.2.3. Comparison between BCTs and CCTs

The load–displacement characteristics of BCTs and CCTs with identical geometrical parameters are depicted in Figure 11. The force–displacement curves for CCTs show a rapid increase in force during the initial impact stage, followed by significant fluctuations around the mean load. These fluctuations correspond to the formation of folds, indicating an unstable energy absorption process. In contrast, the intricate cross-section of the BCT enhances its overall axial impact resistance, resulting in a higher initial peak crush force (*IPCF*). Moreover, BCTs exhibit smaller load fluctuations compared to CCTs, enabling better control over the energy absorption capacity. This controlled deformation results in a more efficient energy absorption process. Further studies indicate that BCTs not only possess a lower ultimate load capacity (*ULC*), but also achieve a similar or even higher specific energy absorption (*SEA*) compared to CCTs, making them more efficient in terms of energy absorption. The PEEQ (equivalent plastic strain) contours illustrated in Figure 12 reveal the irreversible plastic deformation induced by the crushing load, which ultimately determines the energy absorption capacity of the tubes [30]. For CCTs, two folds are observed at 28 mm and 92 mm from the moving flat plate, both occurring symmetrically from the horizontal center of the tube and extending towards the ends. This symmetrical folding pattern indicates a localized deformation, which can lead to higher load fluctuations. These findings suggest that BCTs may be more suitable for applications requiring high energy absorption capacity and stable deformation behavior under compressive loads.

In contrast, BCTs exhibit a progressive buckling deformation pattern along the axial corrugation. Pleats initially form at the tube ends, followed by a progressive deformation along the length of the tube. Subsequently, additional pleats emerge from the horizontal center and extend towards the ends. This progressive buckling allows for a more distributed energy absorption process, minimizing the load fluctuations. Notably, the PEEQ regions are concentrated near the crushing interface, indicating effective energy absorption at these points. The irreversible plastic deformation induced by the crushing load is focused in these regions, ultimately determining the energy absorption capacity of the tubes.

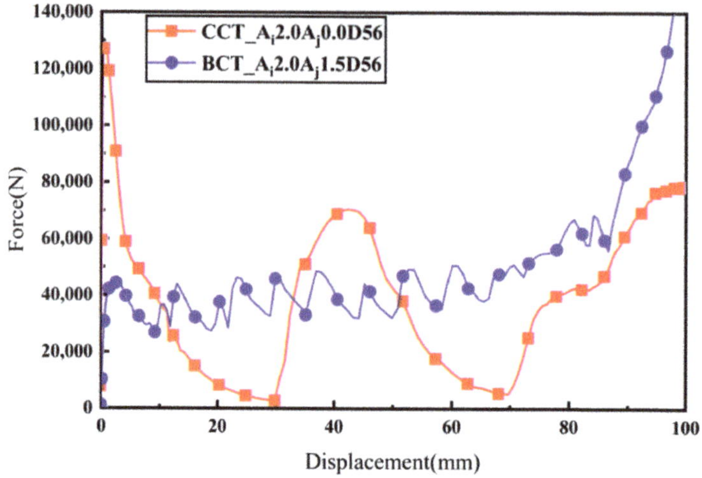

Figure 11. Load–displacement diagrams of the CCT and BCT under axial compressive loading.

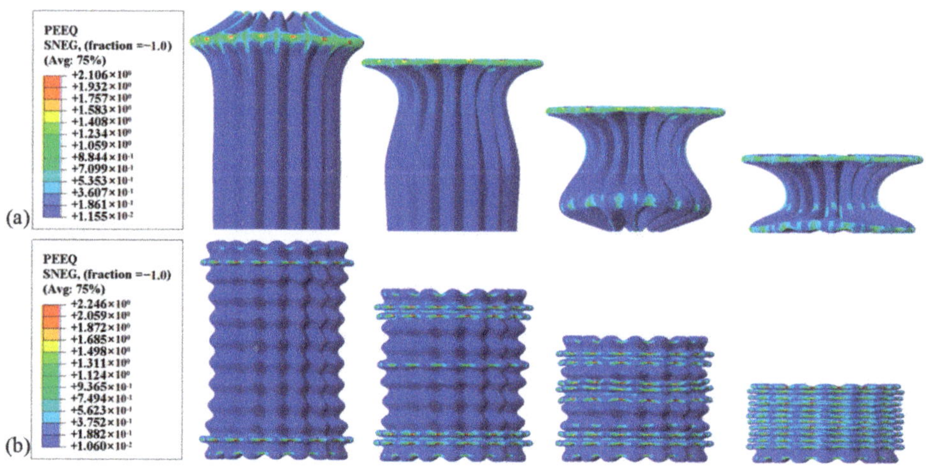

Figure 12. Equivalent plastic strain (PEEQ) contours: (**a**) CCT and (**b**) BCT.

The load–displacement response of BCTs exhibits minimal fluctuations during the crushing process, thereby reducing the risk of injury and enhancing crashworthiness. Additionally, the crushing displacement required to form the second fold in BCTs is significantly less than that in CCTs, yet the energy absorption is substantially greater. This indicates that BCTs are capable of absorbing more energy over a shorter deformation distance, enhancing their efficiency. By reducing the required deformation distance, BCTs can potentially enable the design of lighter and more space-efficient crashworthy structures, which is crucial for applications where weight and space are constrained.

Overall, the force–displacement curves of CCTs exhibit significant fluctuations during axial crushing, consistent with Deng's previous findings [11]. This behavior indicates that CCTs require more energy to develop hinges, which can be attributed to their intricate cross-sectional design that enhances overall structural stiffness. In contrast, BCTs demonstrate a smoother energy absorption pattern due to the axial ripples that dictate the fold locations, allowing the tube to deform in a controlled manner according to the pre-folding, as illustrated in Figure 12.

This figure shows that the specific energy absorption (SEA) of BCTs surpasses that of CCTs by up to 80.23% in the first dataset. This substantial increase in SEA demonstrates that BCTs are more effective at absorbing energy per unit mass compared to CCTs. Simultaneously, the ultimate load capacity (ULC) of BCTs is reduced by 57.63% relative to CCTs. The significant reduction in ULC highlights the BCT's capability to deform under lower peak loads, which can be advantageous in crash scenarios where controlled deformation is crucial. Here, the SEA of BCTs is shown to be only 11.15% higher than that of CCTs. Despite this relatively modest increase in SEA, the BCT achieves a remarkable ULC reduction of up to 72.96% compared to CCTs. This indicates that BCTs offer a more efficient energy absorption profile, with a substantial reduction in peak loads, which could be beneficial in mitigating impact forces during a crash. These panels further illustrate that BCTs not only maintain a high SEA compared to CCTs of the same size but also exhibit a smoother energy absorption curve throughout the crushing process. The smoother energy absorption is attributed to the BCT's design, which allows for more controlled and consistent deformation. The axial ripples in BCTs pre-determine the fold locations, enabling the tube to deform in a predictable and stable manner. This results in reduced fluctuations in load and enhanced performance in energy dissipation.

In summary, Figure 13 highlights that BCTs offer superior energy absorption capabilities compared to CCTs, with a higher SEA and lower ULC. The improved performance of BCTs in energy absorption, combined with their ability to maintain smoother load–displacement behavior, demonstrates their potential for enhanced crashworthiness in practical applications.

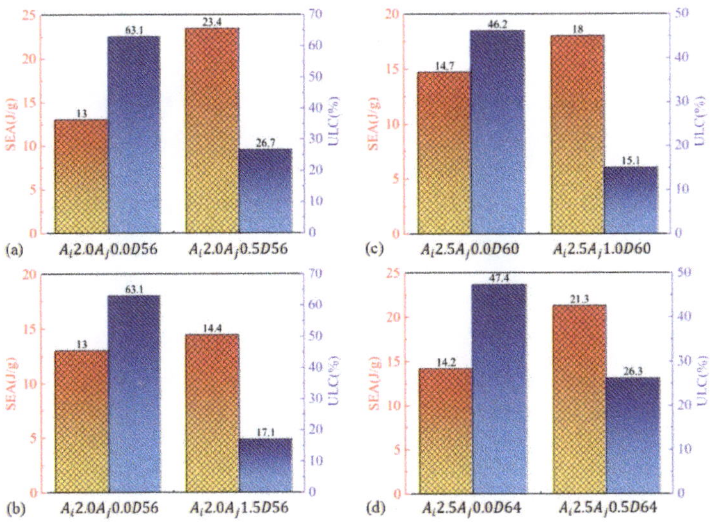

Figure 13. Crashworthiness properties of CCTs and BCTs.

3.2.4. Comparison between BCTs and LCTs

The comparative analysis presented in Figure 14 offers profound insights into the load–displacement characteristics of BCTs in contrast to LCTs with identical geometrical parameters. This comparison reveals several key insights into the mechanical performance of BCTs relative to LCTs. Furthermore, the observation that BCTs can withstand augmented loads while maintaining or even reducing their ULC compared to LCTs underscores a notable enhancement in their load-bearing capabilities. Despite having a comparable or lower peak load, BCTs exhibit a unique advantage in their ability to handle higher loads without experiencing significant failure or instability. This is indicative of the BCTs' efficient load distribution and deformation characteristics. BCTs demonstrate the capacity to withstand higher loads while maintaining or reducing the ULC. This suggests that BCTs offer enhanced load-bearing capabilities and better performance under compressive forces. The improved load resistance of BCTs is likely due to their structural design, which allows for more effective energy absorption and load distribution. The ability of BCTs to maintain similar or lower ULC while withstanding higher loads highlights their potential for applications where both high energy absorption and controlled deformation are required. This makes BCTs suitable for applications such as impact protection and crashworthiness, where efficient energy dissipation and structural resilience are critical. In summary, Figure 14 highlights that BCTs not only match or exceed the performance of LCTs in terms of load-bearing capacity, but also demonstrate the ability to withstand higher loads while maintaining similar or lower ULC values. These findings suggest that BCTs offer significant advantages in terms of mechanical performance and structural efficiency.

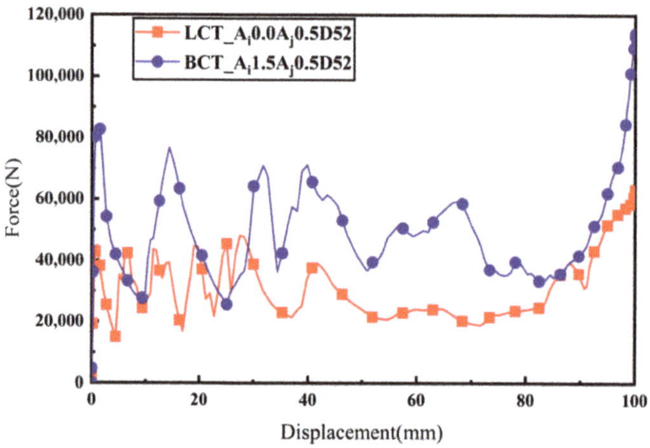

Figure 14. Load–displacement diagrams of the LCT and BCT under axial compressive loading.

Figure 15 presents a comparative analysis of the PEEQ contours for both the LCT and the BCT, revealing distinct deformation behaviors under compressive loading. The PEEQ contours for the LCT show that the initial two folds form at the extremities of the tube, followed by a significant global buckling deformation at the midpoint. This hybrid deformation pattern indicates that the LCT undergoes localized folding at the ends, which then transitions into a global buckling mode. This results in a less efficient energy absorption process, as the deformation is not uniformly distributed along the length of the tube. In contrast, the BCT initially exhibits folding at a distance of 24.54 mm from the moving plate's displacement. Following this initial deformation, the successive layers of the unit experience destabilization, with the upper end of each layer moving outward and the lower end moving inward. This local destabilization leads to a global buckling deformation in the lowest four layers of the BCT. This pattern demonstrates a more progressive and controlled deformation, which contributes to more efficient energy absorption. Throughout the crushing process, the load–displacement response of the LCT shows minimal fluctuation but also low energy absorption. This behavior is indicative of the LCT's less effective energy dissipation, as the hybrid deformation mode results in reduced overall performance in energy absorption. Conversely, the BCT's load–displacement response exhibits more pronounced fluctuations initially, transitioning to a relatively smooth response. This transition reflects the BCT's ability to absorb energy more effectively over a larger deformation range, resulting in a significantly higher energy absorption (EA) compared to the LCT. The presence of circumferential corrugation in the BCT enhances its structural stiffness and contributes to a more efficient deformation mode, improving overall crashworthiness. The observed differences in deformation modes between the LCT and BCT highlight the impact of structural design on crashworthiness. The LCT, with its axial corrugation, undergoes a combination of deformation modes that reduce its energy absorption capacity. In contrast, the BCT's circumferential corrugation improves its structural performance, allowing for more effective energy dissipation and a smoother load–displacement response. These findings underscore the advantages of BCT design in achieving better energy absorption and structural efficiency, consistent with previous studies such as those by Eyvazian et al. [16].

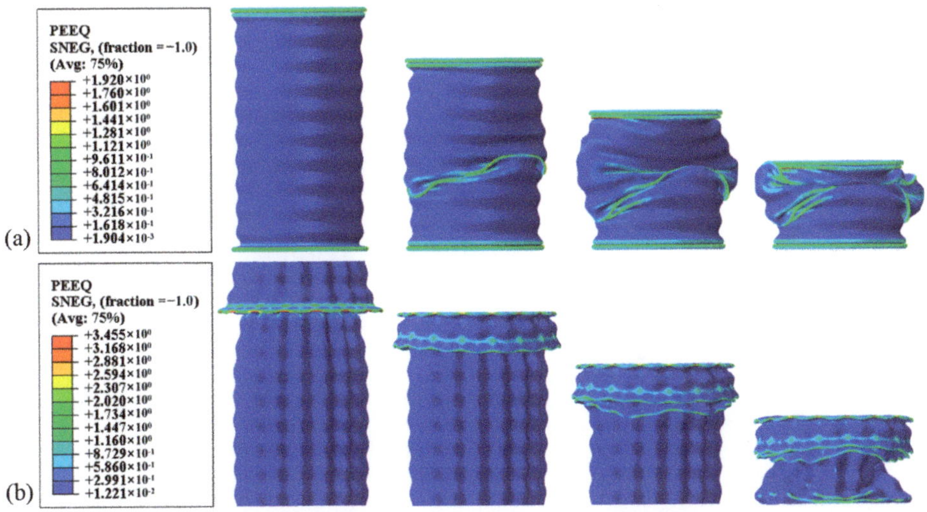

Figure 15. Equivalent plastic strain (PEEQ) contours: (**a**) LCT and (**b**) BCT.

Figure 16 provides a comprehensive comparison of crashworthiness parameters for BCTs and LCTs across four different tube configurations, highlighting the performance differences between these designs. This figure shows that the BCT, with identical geometrical parameters to the LCT, achieves a significant 36.43% reduction in ultimate load capacity (ULC). This significant reduction indicates that the BCT experiences lower peak loads during deformation, which can contribute to reduced risk of catastrophic failure. Additionally, the BCT demonstrates a 36.33% increase in specific energy absorption (SEA), reflecting its superior ability to absorb energy per unit mass. The increase in SEA suggests that the BCT is more effective in dissipating energy during impact, enhancing overall crashworthiness. Here, the BCT exhibits up to 44.9% higher SEA compared to the LCT with the same geometric parameters. This substantial improvement in SEA further underscores the BCT's effectiveness in energy absorption. The BCT also shows a 3.64% reduction in ULC, reinforcing its capability to handle impact loads more efficiently while maintaining a lower peak load. This combined effect of lower peak loads and heightened energy absorption capabilities positions BCTs as a highly favorable structural solution for applications requiring robust crashworthiness performance, which provides additional confirmation of the BCT's superior performance in energy absorption. The BCT consistently maintains a lower ULC and demonstrates higher SEA compared to the LCT of equivalent size. The results indicate that the BCT offers better overall energy absorption capabilities, which is crucial for applications requiring enhanced impact resistance and controlled deformation. The significant increase in SEA for the BCT compared to the LCT indicates a more efficient energy dissipation mechanism. This improvement is vital for applications where maximizing energy absorption is essential for safety and performance. Furthermore, the reduction in ULC observed in the BCT highlights its advantage in experiencing lower peak loads, which can lead to improved structural performance and reduced risk of failure during impact scenarios. The combination of higher SEA and lower ULC confirms that the BCT outperforms the LCT in terms of crashworthiness. This makes the BCT a more effective choice for applications that require superior energy absorption and load management. In summary, Figure 16 conclusively illustrates that the BCT provides substantial improvements in crashworthiness over the LCT, as evidenced by its higher SEA and lower ULC. These performance metrics underscore the BCT's exceptional energy absorption capabilities and overall efficiency in mitigating the effects of impact, making it an attractive option for a wide range of demanding applications.

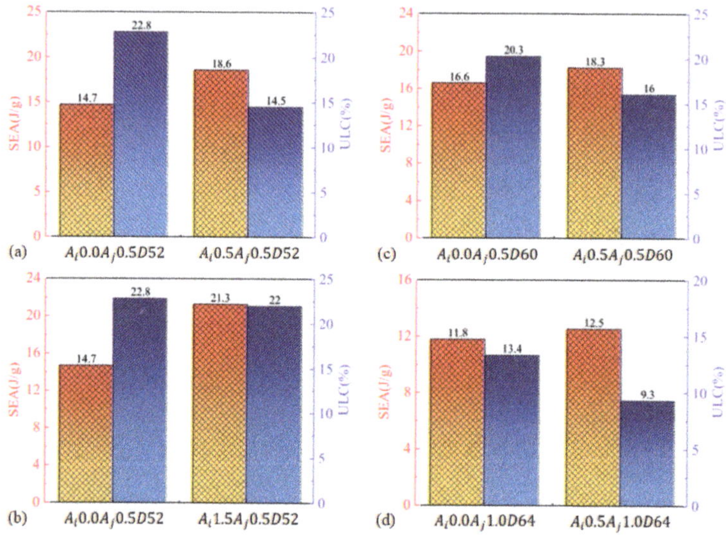

Figure 16. Crashworthiness properties of LCTs and BCTs.

3.3. *Effect of Wall Thickness*

Figure 17a provides a comprehensive illustration of the force–displacement curves for BCTs with varying wall thicknesses ranging from 0.6 to 1.6 mm. As the wall thickness (t) of the BCTs increases, there is a noticeable rise in both the platform force and initial stiffness. This trend indicates that thicker walls contribute to a greater resistance against deformation, resulting in higher peak forces during impact. The increased initial stiffness reflects a more rigid response at the onset of deformation, which is beneficial for applications requiring enhanced structural support. The observed increase in platform force with greater thickness suggests that thicker BCTs can sustain higher loads before yielding or collapsing. This enhancement in load-bearing capacity is critical for applications that demand high impact resistance and durability. Figure 17b presents in a radar plot the crashworthiness parameters of BCTs with different wall thicknesses. ULC remains relatively constant at approximately 12% across different thicknesses. This consistency suggests that the wall thickness has a minimal effect on the maximum load-bearing capacity of the BCTs, implying that other factors may play a more significant role in determining ULC. CFE is around 80% for all thicknesses, indicating a high level of energy absorption efficiency relative to the force applied. This efficiency is maintained across the range of wall thicknesses, demonstrating that the BCT's design effectively manages impact forces regardless of thickness.

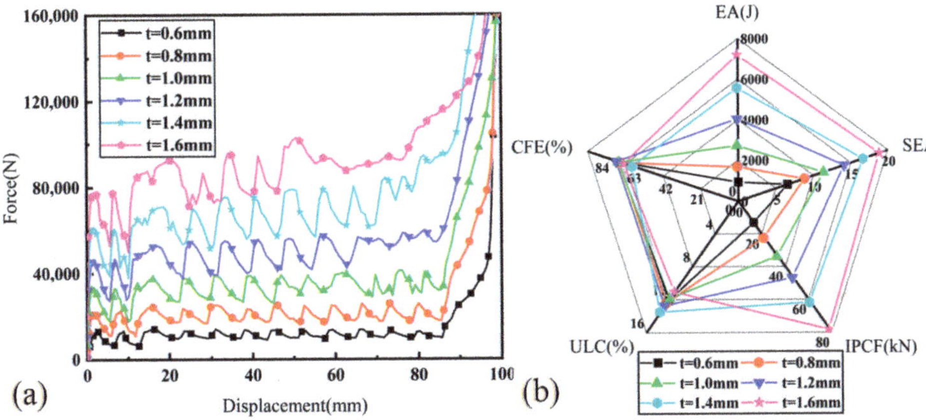

Figure 17. BCT with different thicknesses: (**a**) load—displacement curves; (**b**) radar plot of crashworthiness parameters.

In summary, Figure 17 demonstrates that increasing wall thickness in BCTs enhances their force–displacement response and improves several crashworthiness parameters. While CFE and ULC remain relatively stable, $IPCF$, EA, and SEA benefit significantly from increased thickness. These findings suggest that optimizing wall thickness is crucial for maximizing the impact performance and energy absorption capabilities of BCTs.

4. Optimization

4.1. Definition of the Optimization Problem

To assess the impact of parameters on the crashworthiness of the selected tube, the $N16M12$ configuration was identified as optimal. Variations in the parameter D cause the BCT to transition into the densification stage more rapidly. As an energy absorber, the BCT should maximize the crushing energy per unit mass, making SEA a crucial maximization objective. Additionally, the mechanical response of the energy absorbing structure should exhibit a long and stable stress plateau, making ULC another important metric. A lower ULC indicates a more stable platform force. Thus, the objective is to design a BCT that combines higher SEA with lower ULC. To optimize these crashworthiness metrics, parameters A_i, A_j, D, and t are considered variables for optimization. The goal is to achieve high specific energy absorption while maintaining stable energy absorption. Accordingly, a multi-objective optimization framework was established to meet these criteria:

$$\begin{cases} min(-SEA, ULC) \\ st\ 0\text{ mm} \ll A_i \ll 2\text{ mm} \\ 0\text{ mm} \ll A_j \ll 2\text{ mm} \\ 50\text{ mm} \ll D \ll 60\text{ mm} \\ 0.65\text{ mm} \ll t \ll 1.35\text{ mm} \end{cases} \quad (8)$$

Here, a simultaneous multi-objective optimization of the CCT and LCT was performed to compare and evaluate the crashworthiness performance of the BCT relative to both the CCT and LCT. The process of multi-objective optimization is outlined as follows:

$$\begin{cases} min(-SEA, ULC) \\ st\ 0\text{ mm} \ll A_i \ll 2\text{ mm} \\ 50\text{ mm} \ll D \ll 60\text{ mm} \\ 0.65\text{ mm} \ll t \ll 1.35\text{ mm} \end{cases} \quad (9)$$

$$\begin{cases} min(-SEA, ULC) \\ st\ 0\ \text{mm} \ll A_j \ll 2\ \text{mm} \\ 50\ \text{mm} \ll D \ll 60\ \text{mm} \\ 0.65\ \text{mm} \ll t \ll 1.35\ \text{mm} \end{cases} \tag{10}$$

The optimization process is detailed in Figure 18. Subsequently, sample points are determined using a design-of-experiments approach and subsequently simulated using ABAQUS/Explicit. The resulting simulation data are utilized to develop a surrogate model through the kriging method. Finally, multi-objective optimization (MOO) is conducted using the NSGA-II algorithm to derive the final Pareto frontier.

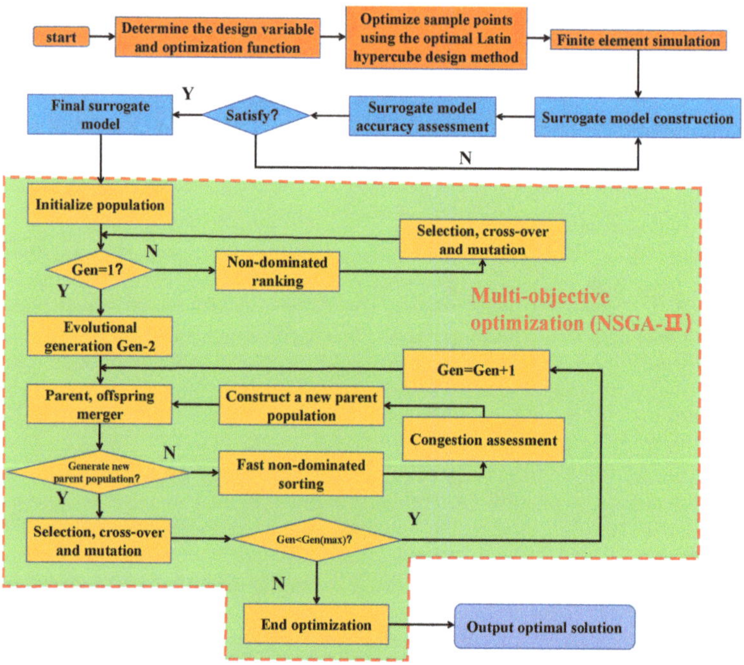

Figure 18. NSGA-II process flow diagram.

4.2. Approximate Model

To manage the extensive experimental data, the optimal Latin hypercube design (OLHD) method is employed for determining the sample points, owing to its excellent space-filling and uniform properties. Using this approach, surrogate models for ULC and SEA are developed via the kriging method. Specifically, 51, 21, and 21 sample points are determined for constructing surrogate models for the BCT, CCT, and LCT, respectively, ensuring model accuracy. The crashworthiness parameters corresponding to these sample points are calculated using the finite element method.

Once the surrogate model is constructed, its accuracy is evaluated using a set of 20 sample points. Four evaluation metrics—coefficient of determination (R^2), relative average absolute error ($RAAE$), relative maximum absolute error ($RMAE$), and root mean square ($RMSE$)—are employed to assess the validity of the constructed model [11]. The specific formulas for these metrics are as follows:

$$R^2 = 1 - \frac{\sum_{i=1}^{n}(y_i - \hat{y}_i)^2}{\sum_{i=1}^{n}(y_i - \overline{y}_i)^2} \tag{11}$$

$$RAAE = \frac{\sum_{i=1}^{n} |y_i - \hat{y}_i|}{n \times \sqrt{\frac{1}{n-1} \sum_{i=1}^{n} (y_i - \hat{y}_i)}} \tag{12}$$

$$RMAE = \frac{max(|y_1 - \hat{y}_1|, |y_2 - \hat{y}_2|, \ldots, |y_n - \hat{y}_n|)}{\sqrt{\frac{1}{n-1} \sum_{i=1}^{n} (y_i - \hat{y}_i)}} \tag{13}$$

$$RMSE = \sqrt{\frac{1}{n} \sum_{i=1}^{n} (y_i - \hat{y}_i)^2} \tag{14}$$

Here, y_i represents the true values, $\hat{y}_i$ denotes the predicted values, $\overline{y}_i$ is the mean value of the sample points, and n is the number of specimens. The model's accuracy improves as the values of $1 - R^2$, $RAAE$, $RMAE$, and $RMSE$ approach 0. Table 3 summarizes the accuracy metrics for the surrogate models of SEA and ULC, demonstrating that the constructed models exhibit high precision.

Table 3. Accuracy evaluation of the surrogate model for BCTs, CCTs, and LCTs.

Columns	Index	R^2	*RAAE*	*RMAE*	*RMSE*
BCTs	*ULC*	0.9119	0.0572	0.2391	0.0924
	SEA	0.9370	0.0662	0.1400	0.0793
CCTs	*ULC*	0.9004	0.1113	0.2176	0.1264
	SEA	0.9357	0.0827	0.1399	0.0917
LCTs	*ULC*	0.9246	0.0887	0.1632	0.1021
	SEA	0.9704	0.0547	0.1047	0.0622

4.3. Optimization Results

The two objectives of the structure are further optimized using the NSGA-II algorithm to derive the Pareto solution set [31]. The optimized Pareto frontier is illustrated in Figure 19. This figure demonstrates that SEA and ULC are conflicting objectives; theoretically, all solutions along the Pareto frontier are optimal and provide valuable guidance for engineers in making decisions based on specific requirements. For instance, point A represents the optimal solution when SEA is prioritized, while point B is ideal for maximizing ULC. The trade-off between SEA and ULC is determined using the minimum distance selection method (TMDSM), which identifies the "knee point" that offers a balanced compromise between the two objectives. The mathematical expression for this compromise point is given by [32]:

$$minD = \sqrt{\frac{1}{n} \sum_{\tau=1}^{n} \left(\frac{f_{c\tau} - min(f_\tau(x))}{max(f_\tau(x)) - min(f_\tau(x))} \right)^2} \tag{15}$$

where $f_{c\tau}$ represents the τth objective value in the Pareto solution for the cth point, Min$D$ is the minimum distance from the "utopia point" to a point on the Pareto frontier, and n is the number of optimization points.

Finally, a comparison between the numerical results and the optimized solutions reveals notable insights, as depicted in Table 4. The maximum discrepancies observed in SEA and ULC between the optimized and numerical results are 2.12% and 3.86%, respectively, indicating a high level of accuracy in the optimized results.

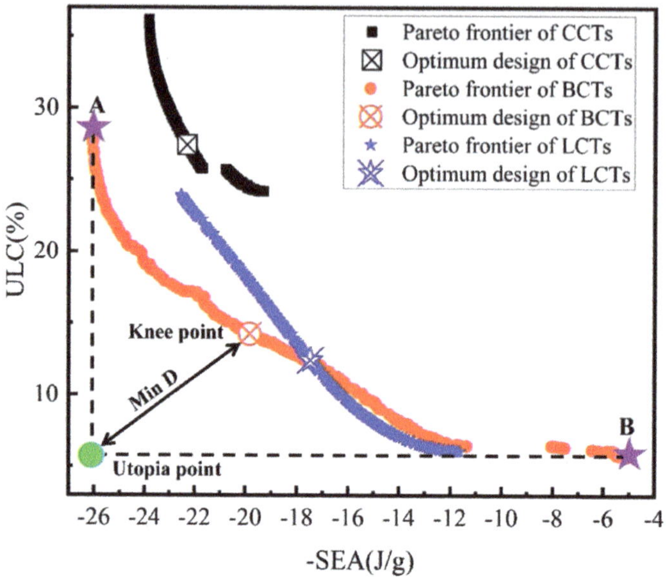

Figure 19. Pareto front of the BCT structure.

Table 4. Comparison of the optimization value and finite element analysis results.

Item	A_i	A_j	D	t	NSGA-II		FEA Results		Accuracies	
					SEA (J/g)	*ULC* (%)	*SEA* (J/g)	*ULC* (%)	RE (%)	RE (%)
Optimal BCTs	1.55	1.19	51.27	1.35	19.80	14.25	20.22	13.70	2.12	3.86

Table 5 presents a comparative analysis of the crashworthiness indices—SEA and ULC—across BCTs, CCTs, and LCTs. When SEA is the primary objective, BCTs demonstrate significant improvements, with increases of 16.72% and 49.14% over optimal CCTs and LCTs designs, respectively. However, these improvements are accompanied by increases in ULC of 3.39% and 126.93%, respectively. Conversely, when minimizing ULC is prioritized, BCTs achieve reductions of 79.09% and 53.94% compared to CCTs and LCTs, respectively, though this results in decreases in SEA of 77.50% and 71.25%, respectively. Notably, the optimal BCT design reduces ULC by 47.99% with only a modest decrease in SEA compared to the optimal CCT design and enhances SEA by 13.40% with a slight increase in ULC compared to the optimal LCT design. These findings highlight the superior crashworthiness of the optimized BCTs through multi-objective optimization.

Table 5. Comparison of crashworthiness data for the CCTs, LCTs, and BCTs optimization solution.

Item	Index	A_i	A_j	D	t	NSGA-II		RE (Compared to CCTs)		RE (Compared to LCTs)	
						SEA (J/g)	*ULC* (%)	*SEA* (J/g)	*ULC* (%)	*SEA* (J/g)	*ULC* (%)
BCTs	Ideal max *SEA*	0.93	0.33	54.01	1.35	26.04	28.23	16.72	3.39	49.14	126.93
	Ideal min *ULC*	0.53	1.86	55.32	0.76	5.02	5.73	−77.50	−79.09	−71.25	−53.94
	Optimal BCTs	1.55	1.19	51.27	1.35	19.80	14.25	−11.25	−47.99	13.40	14.55
CCTs	Optimal CCTs	0.09	0	50.01	1.35	22.31	27.40	-	-	-	-
LCTs	Optimal LCTs	0	0.83	50.00	1.35	17.46	12.44	-	-	-	-

5. Conclusions

To enhance the crashworthiness of thin-walled tubes, this study introduces a novel corrugated structure on both the cross-section and along the wall of the tube, leading to the design of the BCT. The crashworthiness of BCTs was evaluated through axial compression experiments and finite element analysis, yielding the following key conclusions:

1. The BCT is fabricated using a bending method, with the accuracy of the finite element model validated by experimental verification.
2. Discretization analysis assesses the influence of parameters N and M on the crashworthiness of the BCT. The TOPSIS ranking method identifies the optimal values for N and M, which are determined as 12 and 16, respectively.
3. A full-parameter design of experiments investigates the effects of A_i, A_j, and D on structural crashworthiness. A_i influences the energy absorption capacity, A_j enhances the stability of the platform force, and an increase in D improves the likelihood of achieving an optimal energy-absorbing configuration.
4. The SEA of the BCT exceeds that of the CCT by up to 80.23%, while the ULC is reduced by 57.63%. Additionally, the ULC of the BCT is decreased by 72.96% compared to the CCT, and it exhibits a 36.43% lower ULC compared to the LCT, with up to 44.9% higher SEA than the LCT.
5. A simultaneous multi-objective optimization of CCT, LCT, and BCT based on the NSGAII method reveals that the Pareto front of the BCT demonstrates superior crashworthiness advantages compared to both CCT and LCT. The proposed BCT shows significant potential as an energy absorber with promising applications across various fields.

Given its stable reaction load and high SEA value, the BCT holds substantial promise in crashworthiness engineering. In automotive applications, it can be used for the crash box behind the front bumper, and in aerospace, it can be utilized for the bottom plate strut. Further research will focus on analyzing its mechanical properties under high-speed impact and studying its vibration characteristics, ensuring a progressive examination of this work.

Author Contributions: Conceptualization, L.Z. and W.Y.; methodology, X.W.; software, X.H.; validation, L.Z. and M.L.; formal analysis, S.L.; investigation, R.W.; resources, Z.J.; data curation, W.Y.; writing—original draft preparation, L.Z.; writing—review and editing, Z.J.; visualization, S.L.; supervision, W.Y.; project administration, W.Y.; funding acquisition, W.Y. All authors have read and agreed to the published version of the manuscript.

Funding: This research was supported by the Science Foundation of the National Key Laboratory of Science and Technology on advanced composites in special environments (6142905222701).

Institutional Review Board Statement: Not applicable.

Informed Consent Statement: Not applicable.

Data Availability Statement: Data available upon reasonable request.

Conflicts of Interest: The authors declare no conflicts of interest.

Abbreviations

The following abbreviations are used in this manuscript:

LCT	longitudinal corrugated tube
CCT	circumferential corrugated tube
BCT	bi-directional corrugated tube
N	number of units in a single-layer BCT
M	number of layers in the BCT
A_i	amplitude of transverse ripple
A_j	amplitude of axial ripples
D	diameter of the neutral plane of the BCT
t	wall thickness of the BCT
H	height of the BCT
r	distance from the XOZ plane
φ	bending angle
R	half of the outer diameter of the BCT
EA	energy absorption
SEA	specific energy absorption
$IPCF$	initial peak crushing force
MCF	mean crushing force
CFE	crushing force efficiency
ULC	undulation of load carrying capacity
ρ	density
E	Young's modulus
σ_y	yield stress
v	Poisson's ratio
PEEQ	equivalent plastic strain
F_{max}	The maximum value of the load in the load displacement curve

Appendix A

Step one: First create a decision matrix (D_x) and normalize the data for the D_x based on whether the criterion is benefit or cost.

For a problem with n structures and m criteria, D_x is constructed as follows:

$$D_x = \begin{bmatrix} x_{11} & x_{12} & \cdots & x_{1n} \\ x_{21} & x_{22} & \cdots & x_{2n} \\ \vdots & \vdots & \ddots & \vdots \\ x_{m1} & x_{m2} & \cdots & x_{mn} \end{bmatrix} \tag{A1}$$

where x_{ij} is the performance value of the *ith* criterion under the *jth* structure.

The standardized formula is:

$$s_{ij} = \begin{cases} \frac{x_{ij} - min_i(x_{ij})}{max_i(x_{ij}) - min_i(x_{ij})} (criteria\ is\ benefit) \\ \frac{max_i(x_{ij}) - x_{ij}}{max_i(x_{ij}) - min_i(x_{ij})} (criteria\ is\ cost) \end{cases} \tag{A2}$$

where $max_i(x_{ij})$ and $min_i(x_{ij})$ stand for the maximum and minimum values of the samples under the *ith* criterion, respectively.

According to the above equation, the standardized matrix $S = (s_{ij})_{mn}$ can be obtained.

Step two: Normalize each row of the normalized matrix and calculate the entropy value and weights.

The normalized formula is:

$$n_{ij} = \frac{s_{ij}}{\sum_{i=1}^{n} s_{ij}} \tag{A3}$$

The entropy formula for the *ith* standard is:

$$e_i = -\frac{1}{\ln n}\sum_{i=1}^{n} n_{ij}\ln n_{ij} \tag{A4}$$

The weights under each criterion are calculated separately after calculating the entropy value:

$$w_i = \frac{1 - e_i}{\sum_{i=1}^{m}(1 - e_i)} \tag{A5}$$

Step three: Construct a weighted matrix $W = \left(w_{ij}^*\right)_{mn}$ and obtain two ideal solution points.

The weighted matrix is calculated as:

$$w_{ij}^* = w_i \times s_{ij} \tag{A6}$$

The ideal solution points are the maximum and minimum values of under each criterion, respectively:

$$W^{+*} = \left\{w_1^{+*}, w_2^{+*}, \cdots, w_m^{+*}\right\} = \{w_1, w_2, \cdots, w_m\} \tag{A7}$$

$$W^{-*} = \left\{w_1^{-*}, w_2^{-*}, \cdots, w_m^{-*}\right\} = \{0, 0, \cdots, 0\} \tag{A8}$$

Step four: Euclidean distance between each scenario and the ideal scenario with the relative proximity of the alternatives is calculated.

Euclidean distance formulas are, respectively:

$$D_j^+ = \sqrt{\sum_{i=1}^{m}\left(w_i^{+*} - w_{ij}^*\right)^2} = \sqrt{\sum_{i=1}^{m}\left(w_i - w_{ij}^*\right)^2} \tag{A9}$$

$$D_j^- = \sqrt{\sum_{i=1}^{m}\left(w_i^{-*} - w_{ij}^*\right)^2} = \sqrt{\sum_{i=1}^{m}\left(w_{ij}^*\right)^2} \tag{A10}$$

The formula for relative proximity is:

$$Q_j = \frac{D_j^-}{D_j^+ + D_j^-} = \frac{\sqrt{\sum_{i=1}^{m}\left(w_{ij}^*\right)^2}}{\sqrt{\sum_{i=1}^{m}\left(w_i - w_{ij}^*\right)^2} + \sqrt{\sum_{i=1}^{m}\left(w_{ij}^*\right)^2}} \tag{A11}$$

Finally, Q_j is sorted from highest to lowest.

Appendix B

Table A1. Ranking using the entropy TOPSIS method.

Tubes	Q_j	Ranking	Tubes	Q_j	Ranking
N4M6	0.0076	19	N12M10	0.2954	11
N4M8	0.2060	16	N12M12	0.3533	6
N4M10	0.2816	13	N16M6	0.3486	7
N4M12	0.2901	12	N16M8	0.3057	10
N8M6	0.1760	18	N16M10	0.5116	5
N8M8	0.2518	15	N16M12	0.6022	4
N8M10	0.3227	9	N20M8	0.6435	3
N8M12	0.3270	8	N20M10	0.8101	1
N12M6	0.1766	17	N20M12	0.7906	2
N12M8	0.2621	14			

References

1. Zhou, X.W.; Liu, J.D.; Rong, W.F.; Wu, B.Y. Axial Crushing Theory and Optimization of Lattice-Filled Multicellular Square Tubes. *Materials* **2024**, *17*, 1245. [CrossRef]
2. Li, W.W.; Li, Z.H.; Li, S.H.; Wang, P. Crushing Behaviors and Energy Absorption Evaluation Methods of Hexagonal Steel Tubular Columns with Triangular Cells. *Materials* **2022**, *15*, 3910. [CrossRef]
3. Tan, X.; Zhu, S.; Wang, B.; Kadic, M. Tuning negative stiffness mechanical metamaterial's snap-through behavior with a series-connected spring. *Eur. J. Mech. A Solids* **2024**, *107*, 105382. [CrossRef]
4. Zhou, J.; Jia, S.; Qian, J.; Chen, M.; Chen, J. Improving the Buffer Energy Absorption Characteristics of Movable Lander-Numerical and Experimental Studies. *Materials* **2020**, *13*, 3340. [CrossRef]
5. Tan, X.; Cao, B.; Liu, W.; Ji, C.; Wang, B.; Li, S. Odd mechanical metamaterials with simultaneously expanding or contracting under both compression and tension. *Thin-Walled Struct.* **2024**, *203*, 112225. [CrossRef]
6. Xing, J.; Zhao, J.L.; Niu, Q.; Zhang, T.Y.; Zhang, C.Y.; Zhang, Y.L.; Wang, W.Z.; Yan, S.Z.; Liu, X.A. Crashworthiness design and optimization of bamboo-inspired tube with gradient multi-cells. *Thin-Walled Struct.* **2023**, *191*, 111034. [CrossRef]
7. Li, Z.; Chen, W.; Hao, H. Numerical study of open-top truncated pyramid folded structures with interconnected side walls against flatwise crushing. *Thin-Walled Struct.* **2018**, *132*, 537–548. [CrossRef]
8. Liu, Z.L.; Wang, Y.; Liang, X.; Yu, W. Crashworthiness Study of Functional Gradient Lattice-Reinforced Thin-Walled Tubes under Impact Loading. *Materials* **2024**, *17*, 2264. [CrossRef]
9. Tao, C.L.; Zhou, X.; Liu, Z.L.; Liang, X.; Zhou, W.T.; Li, H.J. Crashworthiness Study of 3D Printed Lattice Reinforced Thin-Walled Tube Hybrid Structures. *Materials* **2023**, *16*, 1871. [CrossRef]
10. Wu, S.; Sun, G.; Wu, X.; Li, G.; Li, Q. Crashworthiness analysis and optimization of fourier varying section tubes. *Int. J. Non-Linear Mech.* **2017**, *92*, 41–58. [CrossRef]
11. Deng, X.; Liu, W.; Jin, L. On the crashworthiness analysis and design of a lateral corrugated tube with a sinusoidal cross-section. *Int. J. Mech. Sci.* **2018**, *141*, 330–340. [CrossRef]
12. Deng, X.; Qin, S.; Huang, J. Energy absorption characteristics of axially varying thickness lateral corrugated tubes under axial impact loading. *Thin-Walled Struct.* **2021**, *163*, 107721. [CrossRef]
13. Li, Z.; Yao, S.; Ma, W.; Xu, P.; Che, Q. Energy-absorption characteristics of a circumferentially corrugated square tube with a cosine profile. *Thin-Walled Struct.* **2019**, *135*, 385–399. [CrossRef]
14. Eyvazian, A.; Eltai, E.; Musharavati, F.; Taghipoor, H.; Sebaey, T.A.; Talebizadehsardari, P. Experimental and numerical investigations on axial crushing of square cross-sections tube with vertical wave. *Steel Compos. Struct.* **2020**, *36*, 119–141. [CrossRef]
15. Singace, A.A.; ElSobky, H. Behaviour of axially crushed corrugated tubes. *Int. J. Mech. Sci.* **1997**, *39*, 249–261, 263–265, 267–268. [CrossRef]
16. Eyvazian, A.; Habibi, M.K.; Hamouda, A.M.; Hedayati, R. Axial crushing behavior and energy absorption efficiency of corrugated tubes. *Mater. Des.* **2014**, *54*, 1028–1038. [CrossRef]
17. Dong, J.; Li, W.; Fan, H. Axial crushing behaviors of buckling induced triangular tubular structures. *Mater. Des.* **2021**, *201*, 109513. [CrossRef]
18. Chen, D.H.; Ozaki, S. Numerical study of axially crushed cylindrical tubes with corrugated surface. *Thin-Walled Struct.* **2009**, *47*, 1387–1396. [CrossRef]
19. Eyvazian, A.; Tran, T.N.; Hamouda, A.M. Experimental and theoretical studies on axially crushed corrugated metal tubes. *Int. J. Non-Linear Mech.* **2018**, *101*, 86–94. [CrossRef]
20. Alkhatib, S.E.; Matar, M.S.; Tarlochan, F.; Laban, O.; Mohamed, A.S.; Alqwasmi, N. Deformation modes and crashworthiness energy absorption of sinusoidally corrugated tubes manufactured by direct metal laser sintering. *Eng. Struct.* **2019**, *201*, 109838. [CrossRef]

1. Imai, T.; Shibutani, T.; Matsui, K.; Kumagai, S.; Tran, D.T.; Mu, K.Y.; Maekawa, T. Curvature sensitive analysis of axially compressed cylindrical tubes with corrugated surface using isogeometric analysis and experiment. *Comput. Aided Geom. Des.* **2016**, *49*, 17–30. [CrossRef]
2. Wu, S.; Li, G.; Sun, G.; Wu, X.; Li, Q. Crashworthiness analysis and optimization of sinusoidal corrugation tube. *Thin-Walled Struct.* **2016**, *105*, 121–134. [CrossRef]
3. Yao, S.; Huo, Y.; Yan, K.; Xu, P. Crashworthiness study on circular hybrid corrugated tubes under axial impacts. *Thin-Walled Struct.* **2019**, *145*, 106358. [CrossRef]
4. Ngoc San, H.; Lu, G.; Xiang, X. High energy absorption efficiency of thin-walled conical corrugation tubes mimicking coconut tree configuration. *Int. J. Mech. Sci.* **2018**, *148*, 409–421. [CrossRef]
5. Wang, X.; Yin, W.; Zou, L.; Cai, C.; Li, P.; Zou, H.; Yang, M.; Li, J.; Li, Y. Curved and straight combined cross-section structure for energy absorption. *Int. J. Mech. Sci.* **2022**, *224*, 107333. [CrossRef]
6. Liu, Z.; Hao, W.; Xie, J.; Lu, J.; Huang, R.; Wang, Z. Axial-impact buckling modes and energy absorption properties of thin-walled corrugated tubes with sinusoidal patterns. *Thin-Walled Struct.* **2015**, *94*, 410–423. [CrossRef]
7. Behzadian, M.; Otaghsara, S.K.; Yazdani, M.; Ignatius, J. A state-of the-art survey of TOPSIS applications. *Expert Syst. Appl.* **2012**, *39*, 13051–13069. [CrossRef]
8. Li, Z.; Ma, W.; Yao, S.; Xu, P. Crashworthiness performance of corrugation- reinforced multicell tubular structures. *Int. J. Mech. Sci.* **2021**, *190*, 106038. [CrossRef]
9. Liu, S.; Tong, Z.; Tang, Z.; Liu, Y.; Zhang, Z. Bionic design modification of non-convex multi-corner thin-walled columns for improving energy absorption through adding bulkheads. *Thin-Walled Struct.* **2015**, *88*, 70–81. [CrossRef]
0. Mozafari, H.; Lin, S.; Tsui, G.C.P.; Gu, L. Controllable energy absorption of double sided corrugated tubes under axial crushing. *Compos. Part B Eng.* **2018**, *134*, 9–17. [CrossRef]
1. Xu, F.; Yu, K.; Hua, L. In-plane dynamic response and multi-objective optimization of negative Poisson's ratio (NPR) honeycomb structures with sinusoidal curve. *Compos. Struct.* **2021**, *269*, 114018. [CrossRef]
2. Chen, Y.; Liu, G.; Zhang, Z.; Hou, S. Integrated design technique for materials and structures of vehicle body under crash safety considerations. *Struct. Multidiscip. Optim.* **2017**, *56*, 455–472. [CrossRef]

materials

MDPI

Article

Study on the Approach to Obtaining Mechanical Properties Using Digital Image Correlation Technology

Shuai Wang [1,2,3], Bin Wang [2,*], Shengyong Mu [4], Jianlong Zhang [4], Yubiao Zhang [3] and Xiaoyan Gong [3]

[1] College of Mechanical Engineering, Xi'an Shiyou University, Xi'an 710065, China; shuai.wang.ac@outlook.com
[2] Department of Mechanical and Aerospace Engineering, Brunel University of London, Uxbridge UB8 3PH, UK
[3] School of Mechanical Engineering, Xi'an University of Science and Technology, Xi'an 710054, China; zhangyubiao@stu.xust.edu.cn (Y.Z.); gongxymail@163.com (X.G.)
[4] R&D Department, Xi'an Special Equipment Inspection Institute, Xi'an 710065, China; mushengyong@xseii.com.cn (S.M.); zhangjl@chd.edu.cn (J.Z.)
* Correspondence: bin.wang@brunel.ac.uk

Abstract: Accurate mechanical property parameters constitute an indispensable guarantee for the accuracy of finite element simulations. Traditionally, uniaxial tensile tests are instrumental in acquiring the stress–strain data of materials during elongation, thereby facilitating the determination of the materials' mechanical property parameters. By capitalizing on the digital image correlation (DIC) non-contact optical measurement technique, the entire test can be comprehensively documented using high-speed cameras. Subsequently, through in-depth analysis and meticulous numerical computations enabled by computer vision technology, the complete strain evolution of the specimen throughout the test can be precisely obtained. In this study, a comparison was made between the application of strain gauges and DIC testing systems for measuring the strain alterations during the tensile testing of 316L stainless steel, which serves as the material for the primary circuit pipelines of pressurized water reactor (PWR) nuclear power plants (NPPs). The data procured from these two methods were utilized as material mechanical parameters for finite element simulations, and a numerical simulation of the uniaxial tensile test was executed. The results reveal that, within the measuring range of the strain gauge, the DIC method generates measurement outcomes that are virtually identical to those obtained by strain gauges. Given its wider measurement range, the DIC method can be effectively adopted in the process of obtaining material mechanical parameters for finite element simulations.

Keywords: uniaxial tensile test; elastoplastic finite element method; mechanical properties

1. Introduction

The finite element method plays a vital role in engineering design and simulation. Precise material mechanical properties are among the key factors that ensure the accuracy of finite element simulation results. Material mechanical parameters are commonly determined using uniaxial tensile tests, with accurate measurements of stress–strain data being paramount. When it comes to strain measurement, traditional methods like extensometers and strain gauges have their drawbacks, including a limited measuring range, significant installation errors of measuring tools, and an inability to capture the features of material necking [1,2]. The digital image correlation method is a non-contact, high-precision measurement technique that employs computer vision technology for image processing and numerical calculations to obtain full-field deformation and strain fields at different scales [3,4]. This approach can be applied to measurement areas ranging from very small to large, and its results are easily comparable to finite element results or strain gauge measurements. The DIC method, a technique first proposed in the 1980s, is widely used in laser speckle metrology for surface displacement measurement and its use has witnessed substantial growth, thanks to the pioneering efforts of scholars and researchers in the field. This progress encompasses various dimensions, including the selection of functions, the enhancement of speckle quality, and the development of interpolation algorithms. These advancements have led to significant improvements in DIC's computational accuracy and speed in measuring displacements [5–8]. In recent years, the focus of DIC research has shifted primarily towards broadening its range of applications. Researchers have employed DIC to measure the Young's modulus of aluminum alloys, validating its precision through comparisons with strain gauge measurements [9,10]. By harnessing DIC's full-field measurement capabilities, scholars have explored the impact of factors such as specimen size on the Young's modulus. Additionally, DIC has proven valuable in assessing Poisson's ratio, anisotropic plastic ratio parameters, and flow curves. It has also found applications in high-temperature tensile experiments [11]. Behrad successfully employed experimental DIC methods to predict the stress–strain response of 304 stainless steel specimens across a temperature range spanning from room temperature to 900 °C [12]. Shen et al. combined finite element analysis with DIC measurements to investigate the uniaxial tensile properties of aerospace composite materials at elevated temperatures, highlighting the consistency between finite element predictions and DIC experiments [13]. In addition to tensile mechanical properties, digital image correlation (DIC), combined with complementary testing techniques such as acoustic emission (AE), X-ray computed tomography (X-CT), and scanning electron microscopy (SEM), proves invaluable in scrutinizing residual stresses, damage progression, and crack propagation [14–16]. Zhu et al. utilized miniature ring-core cutting alongside DIC techniques to ascertain residual interfacial stresses in thermal barrier coatings [17]. Zhang et al. integrated DIC with scanning electron microscopy to analyze the deformation and fracture behaviors of pure titanium under various loading conditions, including uniaxial, notched tension, and shear loading [18]. DIC, extending beyond its non-contact, non-destructive, and full-field measurement capacities, facilitates localized displacement measurements and provides in situ monitoring capabilities. The technique has proven highly effective in characterizing the local mechanical properties of non-uniform materials like welded joints and in crack detection [19–21]. In summary, DIC surpasses traditional strain measurement methods, offering clear advantages. The experimental data derived through DIC plays a pivotal role in supporting advancements in material damage assessment, fracture mechanics, and finite element methods. Recent studies by Oka et al. and Novich et al. highlight DIC's potential for nuclear-grade steels [22,23], but systematic validation for 316L under reactor-relevant conditions remains lacking.

In this study, numerical analyses and experiments were carried out to investigate the process of strain variation during uniaxial tensile tests, and the results can help guide the accurate acquisition of mechanical property parameters (curves) required during finite element simulation.

2. Experiment and Procedure

To accurately capture the strain variations during uniaxial tensile testing of materials, this study employed two distinct methods: strain gauges and digital image correlation (DIC) techniques. The strain distribution was assessed using these methods, while a uniaxial tensile testing machine was utilized to apply the loading load. 316L nuclear-grade austenitic stainless steel has been used as the material of the primary circuit pipeline of pressurized water reactor (PWR) nuclear power plants (NPPs) due to its strength and good corrosion resistance [24]. Within the framework of uniaxial tensile experiments, the assessment of material strain was carried out by means of both strain gauge measurements and DIC techniques. For each of these methodologies, three independent experimental trials were meticulously conducted. This comprehensive experimental approach ultimately led to the generation of a cumulative total of six datasets. The stress–strain changes experienced by 316L austenitic stainless steel during the uniaxial tensile experiment were analyzed by using numerical simulation methods and compared with the experimental results obtained.

2.1. Specimen

In accordance with GB/T 228.1-2010, metallic materials are subjected to tensile testing at room temperature [25]. By taking into consideration the type of testing machine fixtures, a stainless steel sheet is meticulously crafted into a sheet-shaped tensile specimen, as illustrated in Figure 1, through the process of wire cutting. The gauge section of the specimen exhibits a section size of 3 mm × 2 mm and a length of 20 mm, following corresponding test standards of GB/T 2039-1997 [26]. Furthermore, the chamfer radius between the clamping section and the gauge section is precisely measured at 8 mm. To ensure that the surface roughness of the specimen remains below 3.2 μm, the surface is polished until a distinct metallic luster becomes apparent.

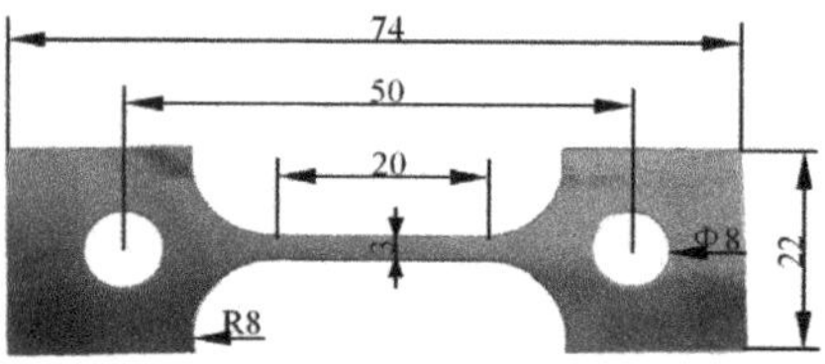

Figure 1. The plate tensile specimen geometric dimensioning (unit: mm).

2.2. Utilizing Strain Gauges to Acquire Strain Data During Material Stretching

Resistance strain gauges are strain measurement components that operate based on the strain effect. The fundamental principle involves the mechanical deformation of strain gauges fixed on a structure when subjected to external loads, leading to changes in their resistance values. Through appropriate calculations, the strain variations in the structure can be determined. The specific experimental procedures for measuring the stress–strain curve of materials using a uniaxial tensile testing machine in conjunction with strain gauges involve the following steps:

(1) Strain gauge patch

Prior to pasting, the gauge section of the specimen is subjected to crosswise polishing along a 45° direction using fine sandpaper to enhance the bonding strength. Following polishing, the surface of the specimen is meticulously scrubbed with cotton yarn. In this particular test, acrylic epoxy resin adhesive is chosen as the adhesive material, and strain gauges are carefully affixed to the surface of the gauge section. The pertinent parameters of the resistance strain gauge BX120-3AA (Telesky, Shenzhen, China) employed in this study are comprehensively presented in Table 1. The parameters in Table 1 are based on the manufacturer's specifications for the strain gauges used in this study.

Table 1. Parameters of strain gauge.

Substrate Material	Wire Material	Length and Width (mm)	Resistance Ω	Sensitivity Coefficient	Nominal Voltage
acetal	constantan	6.5 × 3.5	120.0 ± 0.3	2.06 ± 0.01	≤12 V

(2) Connect strain gauge to deformeter

To facilitate the collection of strain data, the wires of the strain gauges are connected to the extension lines of the strain gauges through welding. This is achieved by utilizing a DH3818Y static strain tester (DONGHUA, Taizhou, China), which allows for the precise and accurate measurement of strain.

(3) Experiment process

The stretching rate of the tensile test is set to 2 mm/min. Following the configuration of the pertinent parameters within the strain acquisition software, both uniaxial tensile testing and strain data acquisition are executed simultaneously. Upon fracture of the specimen, the experiment is terminated. This experiment is carried out at ambient temperature conditions.

2.3. Utilizing DIC to Acquire Strain Data During Material Stretching

The fundamental principle of digital image correlation (DIC) involves the use of artificial speckle fields that are randomly or pseudo-randomly distributed on the surface of a test piece as carriers of deformation information. This enables the experimental mechanics method of full-field displacement and strain analysis of the surface of materials or structures under external load. During DIC testing, a three-dimensional adjustment bracket is utilized to adjust the position of speckle capture and control the leveling of the CCD camera. The light source is primarily used to regulate the intensity and convergence of light on the surface of the sample, resulting in a uniform light field. The recorded images of speckle variations on the sample's surface are then analyzed to extract valuable information about the material's mechanical behavior. The specific steps for measuring strain changes during the uniaxial tensile test of 316L stainless steel using DIC are as follows:

(1) Speckle spraying

In this study, 316L stainless steel specimens of the same batch and size as those used in uniaxial tensile tests and strain gauge measurements were employed. The surface of each specimen was cleaned before being spray-painted to create a uniform speckle pattern with controlled paint concentration, as shown in Figure 2.

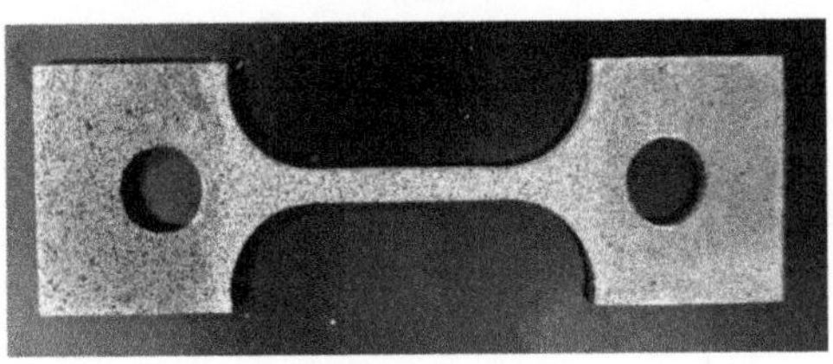

Figure 2. The plate tensile specimen following speckle spray treatment.

After speckle spray treatment, the specimens were vertically clamped onto the fixtures of the loading device. As shown in Figure 3a, the digital image correlation (DIC) measurement system used in this study is a VIC-2D system, which consists of a CCD camera, a light source, and a three-dimensional adjustment stand with a leveler. The data acquisition software used was VIC-Snap 8. An MTS-LPS.105 testing machine (MTS Systems Corporation, Eden Prairie, MN, USA) was employed as the material loading device, with a maximum load capacity of 100 kN. The clamping positions of the specimens are illustrated in Figure 3b.

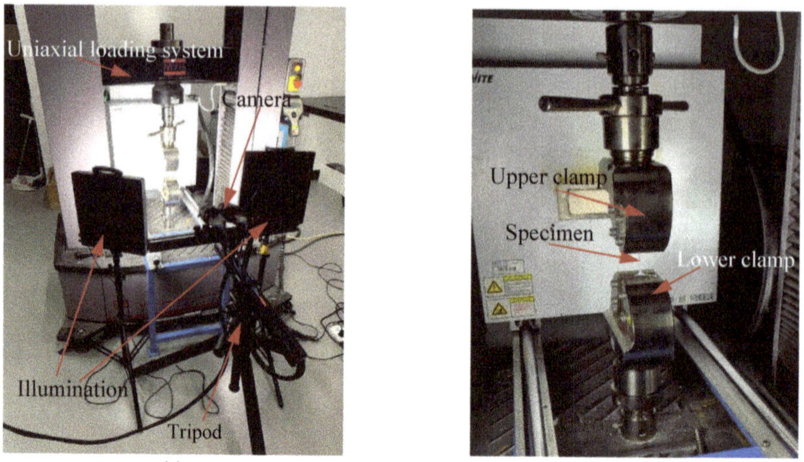

(**a**) (**b**)

Figure 3. Experiment set-up of the tensile testing machine with DIC system. (**a**) DIC system composition. (**b**) The installation position of the specimen.

(2) DIC device parameter adjustment

A CCD camera was mounted on a three-dimensional adjustment stand and aligned parallel to the ground using a leveler. The distances between the CCD camera, the light source, and the specimen were determined to ensure that the deformation of the gauge section of the specimen could be captured by the acquisition system. The light source was turned on to create a uniform light field on the surface of the specimen. The acquisition software was then run, and the aperture was adjusted to an appropriate position on the computer screen before adjusting the focal length until a clear and well-lit image appeared on the acquisition software interface.

(3) DIC software acquisition parameters and tensile speed settings.

The acquisition software (GOM Correlate, version 2020) was set to capture images at a frame rate of two frames per second. Prior to the start of the test, a photograph was taken at the unloaded state to establish a reference position for later software strain calculations. After setting the reference position, the uniaxial tensile test and image acquisition by the CCD camera were simultaneously initiated with a tensile speed of 2 mm/min using the testing machine.

3. Numerical Modeling

In this study, the ABAQUS software (version 2020) was used to complete the finite element analysis process. The FE model is shown in Figure 4, based on the dimensions of the tensile test specimen in Figure 1. For the boundary conditions, we set two reference points on the geometric model of the specimen. The left reference point (RP1) is fixed, and a displacement loading in the X-direction is applied to the center of the right reference point (RP2). The reference points (RP1 and RP2) couple with the inner surface of the loading hole of the specimen and the degrees of freedom outside the X direction are constrained to avoid possible rigid body motion. The code provided element C3D8 (8-node linear brick) was chosen. A total of 1176 elements were found with good convergence to catch the strain distribution on the gauge length of the specimen.

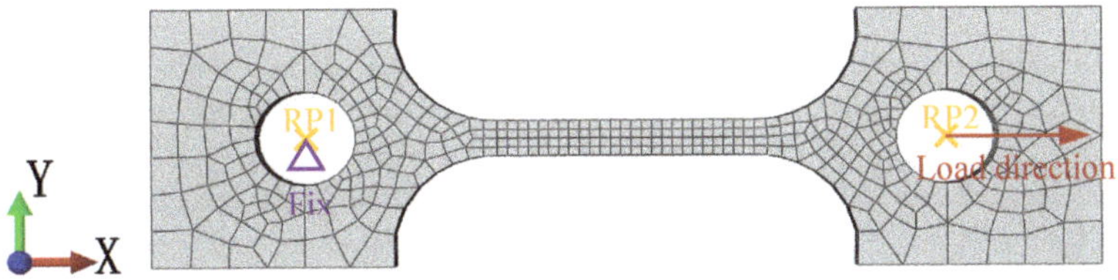

Figure 4. Mesh and boundary condition used for plate tensile specimen.

4. Results and Discussions

4.1. Strain Gauge Test Results

The engineering stress–strain curve ($\sigma_{\text{eng}} - \varepsilon_{\text{eng}}$) of a material can be obtained through uniaxial tensile testing, and then the plastic parameters of the material can be obtained. It should be noted that the true strain ($\varepsilon_{\text{true}}$) of the material is measured by the strain gauge, so it is necessary to convert the engineering stress (σ_{eng}) obtained from the tensile machine to obtain the true stress–strain curve ($\sigma_{\text{true}} - \varepsilon_{\text{true}}$) of the material. The conversion relationship between the engineering stress–strain curve and the true stress–strain curve ($\sigma_{\text{true}} - \varepsilon_{\text{true}}$) is as follows:

$$\sigma_{\text{true}} = \sigma_{\text{eng}}(1 + \varepsilon_{\text{eng}}) \tag{1}$$

$$\varepsilon_{\text{true}} = \ln(1 + \varepsilon_{\text{eng}}) \tag{2}$$

The true stress–strain curve of the material, as measured, is depicted in Figure 5, where the green curve represents the true stress–strain curve of 316L stainless steel obtained through a strain gauge. It can be observed that due to the range limitation caused by the measuring principle of the strain gauge, the strain gauge fails when the strain value in the gauge section reaches approximately 0.05 during the testing process, causing a discontinuity in the curve and rendering it impossible to continue measuring the mechanical property curve of the material. During numerical simulations, significant deformation occurs at locations of stress concentration. To ensure the reliability of the numerical simulation results, a strain data measurement method with a wider range needs to be employed during testing. Notably, all experimental groups demonstrated mechanical response consistency within 2% relative standard deviation, with graphical presentation optimized through the selection of representative datasets from strain gauge and DIC measurement systems.

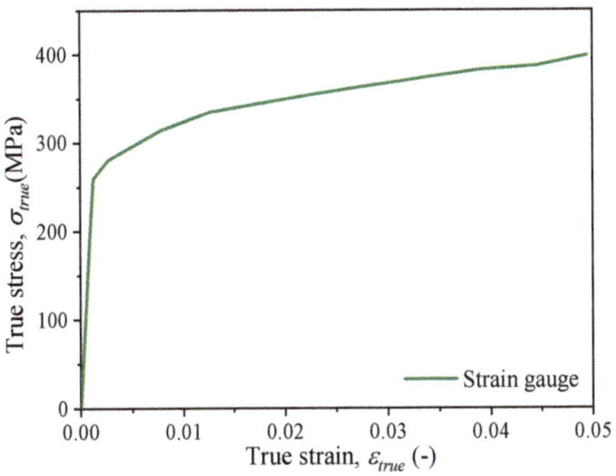

Figure 5. The true stress–strain curve from strain gauge.

4.2. DIC Results Comparison

To obtain the corresponding true stress–strain curve, the engineering stress (σ_{eng}) from the tensile testing machine was converted to true stress (σ_{true}) using Equations (1) and (2). Subsequently, a comparison between the stress–strain curves obtained by the resistance strain gauge and DIC method is shown in Figure 6. A magnified inset was incorporated to highlight the measurement domain of the strain gauge, facilitating the precise observation of strain localization characteristics within the elastic deformation regime. The results demonstrate that within the measurement range of the resistance strain gauge, both methods yield consistent true stress–strain curves for the material, thereby validating the reliability of the DIC method. Minor discrepancies observed in the initial stage may originate from specimen manufacturing tolerances and slight variations in strain gauge bonding accuracy.

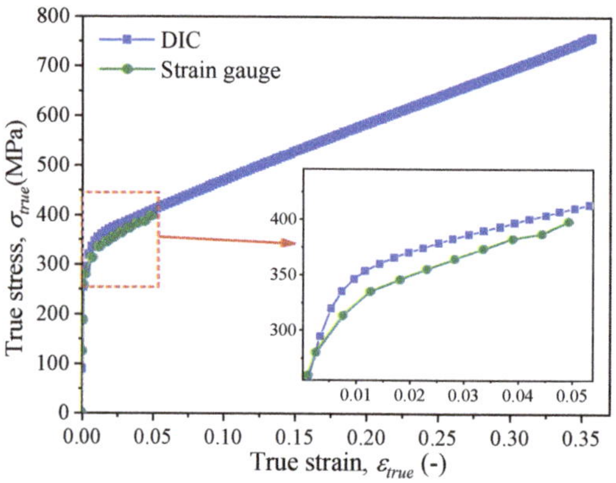

Figure 6. Comparison of the true stress–strain curve from DIC and stain gauge.

The comparison between the two methods reveals that the measurement range of resistive strain gauges is relatively limited, and they cannot function properly when the strain exceeds approximately 0.05. Consequently, resistive strain gauges are unable to obtain full-field strain data from the tested sample. In contrast, non-contact optical measurement methods, such as DIC, can reliably output full-field strain data from the tested sample with a measurement range that extends up to the point of fracture, which is significantly higher than that of strain gauges. Therefore, during numerical simulations, it is feasible to select the true stress–strain data obtained through DIC as the material mechanical property parameters in the finite element model.

The strain cloud diagram of the gauge section of the sample during the tensile process can be obtained through processing using GOM Correlate software, as shown in Figure 7a–c. Since the lower clamp of the tensile testing machine remains fixed, the load is applied through displacement of the upper clamp. Therefore, the symbol D in Figure 7 represents the displacement of the specimen at the gripping location of the upper clamp. From the figures, it can be observed that with the continuous increase in load, the strain value of the gauge section also increases. During the uniaxial tensile test, the strain distribution of the gauge section is relatively uniform, and data on the variation in strain values of the gauge section with time can be outputted through the software. The computational results of the finite element models with different material parameters are shown in Figure 7d–i. Among them, Figure 7d–f shows the strain distribution obtained through finite element calculation using DIC test results as the material mechanical property parameters, while Figure 7g–i uses the strain gauge test results as the material mechanical property parameters. From the figures, it can be observed that when the material undergoes significant plastic deformation, the computational results in Figure 7d–f are generally consistent with the experimental (DIC) results. However, due to limited data based on strain gauge measurements, the finite element computation results deviate from the experimental results at higher levels of plastic deformation. The DIC technique enables full-field strain measurements (as demonstrated in Figure 6), capable of resolving strain distributions in regions inaccessible to conventional strain gauges (e.g., high-strain gradient zones), though its implementation is hindered by the high cost of speckle-recording equipment, limiting accessibility in budget-constrained settings.

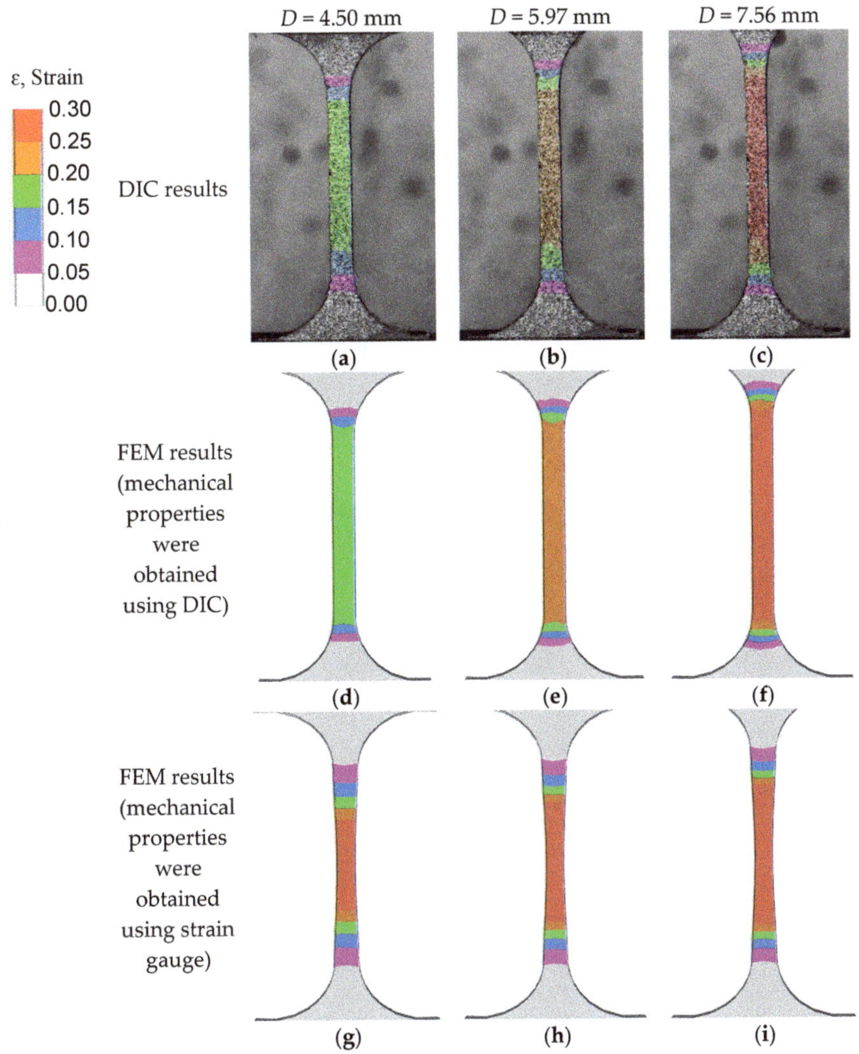

Figure 7. The changes in strain with different mechanical parameters.(**a**) D = 4.50 mm (DIC result); (**b**) D = 5.97 mm (DIC result); (**c**) D = 7.56 mm (DIC result); (**d**) D = 4.50 mm (FFM result using DIC data); (**e**) D = 5.97 mm (FFM result using DIC data); (**f**) D = 7.56 mm (FFM result using DIC data); (**g**) D = 4.50 mm (FFM result using strain gauge data); (**h**) D = 5.97 mm (FFM result using strain gauge data); (**i**) D = 7.56 mm (FFM result using strain gauge data).

The accuracy of DIC-measured strains aligns with recent studies on nuclear materials. For example, Zhang et al. demonstrated that DIC achieves ±0.1% strain accuracy for austenitic steels under quasi-static loading, consistent with our findings [16]. During necking, our observed strain localization matches the patterns in 316L steel reported by Jordan et al., who combined DIC with finite element modeling [27].

5. Conclusions

The following conclusions can be drawn from this study:

(1) When the sample under study experiences significant deformation, the limited measurement range of the strain gauge can impose constraints on the quantity of strain data that can be collected. This, in turn, has a substantial impact on the acquisition of accurate mechanical property parameters. The restricted gauge range acts as a bottleneck, hindering the comprehensive characterization of the material's mechanical behavior during large-scale deformation processes.

(2) The utilization of the DIC technique for strain measurement offers a distinct advantage in enabling precise strain testing over a significantly broader range. This broader measurement scope allows for the collection of more comprehensive material mechanical parameters, which are crucial for conducting accurate elastoplastic finite element simulations. The DIC method provides a more holistic view of the material's strain evolution, thereby enhancing the reliability and accuracy of the finite element models.

(3) On the premise of accurately determining the mechanical property parameters of the material, the finite element simulation-derived strain distribution of 316L steel has been found to be in excellent agreement with the strain distribution obtained through DIC measurements. This congruence validates the effectiveness of the DIC method as an essential validation tool for finite element simulations. The DIC technique not only provides reliable experimental data but also serves as a benchmark for validating the numerical results of finite element models, thereby contributing to the overall improvement of the simulation accuracy and the understanding of the material's mechanical behavior.

For the purpose of structural integrity analyses of key nuclear structures such as reactor pressure vessels (RPVs) and primary circuit pipelines, many scholars and research institutions are making efforts to predict the mechanical properties after long-term service. The validated accuracy of DIC in characterizing homogeneous 316L stainless steel strongly supports its potential application in quantifying the mechanical behavior of heterogeneous nuclear structures, such as welded joints or defect-containing components—critical yet understudied scenarios in primary circuit pipelines. Building on our methodology, where DIC achieved relatively high strain resolution on polished specimens, the technique could map strain localization near weld seams or crack tips in irradiated/welded 316L, overcoming the spatial limitations of strain gauges. For instance, our speckle preparation protocol could be adapted to create controlled artificial defects (e.g., micro-notches mimicking stress corrosion cracks) to study strain redistribution during damage progression. The FE framework, currently calibrated using homogeneous material data, could integrate DIC-derived heterogeneous strain fields to refine constitutive models for weld zones, where microstructure variations (e.g., heat-affected zone and base metal) induce mechanical property gradients. Further study has been planned along this line.

Author Contributions: Conceptualization, B.W.; methodology, S.W.; software, S.M. and Y.Z.; formal analysis, J.Z.; investigation, Y.Z.; resources, X.G.; data curation, S.M. and J.Z.; writing—original draft, S.W.; writing–review and editing, B.W.; supervision, X.G.; funding acquisition, S.W. All authors have read and agreed to the published version of the manuscript.

Funding: This work was financially supported by the State Administration for Market Regulation technical support project, China (2023YJ25); the Natural Science Foundation of Shaanxi province of China (2023-JC-YB-422); the Xi'an Jiaotong University State Key Lab for Strength and Vibration for Mechanical Structures Open Lab Project, China (SV2023-KF-10); Royal Society, UK (IEC\NSFC\233524).

Data Availability Statement: The original contributions presented in this study are included in the article. Further inquiries can be directed to the corresponding author.

Conflicts of Interest: The authors declare no conflict of interest.

References

1. Dong, Z.; Xie, X.; Jiang, W.; Niu, R.; Wan, Y.; Zhai, X.; Zhao, X. Fatigue failure mechanism of duplex stainless steel welded joints including role of heterogeneous cyclic hardening/softening: Experimental and modeling. *Int. J. Fatigue* **2024**, *178*, 108010. [CrossRef]
2. Luo, P.; Li, D.; Zhang, C.; Ru, W.; Han, Z.; Ma, J. Experimental investigation on biaxial mechanical properties of laminated rock and floor heave mechanism of tunnel based on DIC. *Eng. Fail. Anal.* **2024**, *156*, 107848. [CrossRef]
3. Wang, S.; Hou, C.; Wang, B.; Wu, G.; Fan, X.; Xue, H. Mechanical responses of L450 steel under biaxial loading in the presence of the stress discontinuity. *Int. J. Press. Vessel. Pip.* **2022**, *198*, 104662. [CrossRef]
4. Zhang, X.W.; Wen, J.F.; Wang, T.; Zhang, X.-C.; Tu, S.-T. Ductile failure of flat plates containing two through-wall cracks: Experimental investigation and numerical modeling. *Int. J. Press. Vessel. Pip.* **2021**, *191*, 104377. [CrossRef]
5. Obid, Š.; Starman, B.; Urevc, J.; Halilovič, M. Extraction of uniaxial stress–strain curve from bending test using DIC measurements. *Eur. J. Mech. A/Solids* **2024**, *106*, 105346. [CrossRef]
6. Yang, R.; Li, Y.; Zeng, D.; Guo, P. Deep DIC: Deep learning-based digital image correlation for end-to-end displacement and strain measurement. *J. Mater. Process. Technol.* **2022**, *302*, 117474. [CrossRef]
7. Zhang, S.; Liu, X.; Zhou, X.; Wang, X.; Qiu, L.; Gu, Z.; Xie, H. Spatial-temporal response and precursor characteristics of tensile failure on disc coal samples of different sizes combining AE, EMR and DIC techniques. *Eng. Fract. Mech.* **2024**, *309*, 110418. [CrossRef]
8. Zhang, T.; Ma, X.; Yang, B.; Pei, X.; Ge, Z.; Jiang, W. An adaptive spherical indentation test integrating targeted testing scenarios, data acquisition, and model selection for uniaxial mechanical property predictions. *Int. J. Press. Vessel. Pip.* **2024**, *212*, 105365. [CrossRef]
9. Venkatachalam, S.; Banjare, R.; Murthy, H.; Rao, B.C. Mechanical testing of micro-specimens of Al6061-T6 using DIC for strain measurement. *Exp. Tech.* **2019**, *43*, 125–135. [CrossRef]
10. Quanjin, M.; Rejab, M.R.M.; Halim, Q.; Merzuki, M.; Darus, M. Experimental investigation of the tensile test using digital image correlation (DIC) method. *Mater. Today Proc.* **2020**, *27*, 757–763. [CrossRef]
11. Nguyen, V.T.; Kwon, S.J.; Kwon, O.H.; Kim, Y.-S. Mechanical properties identification of sheet metals by 2D-digital image correlation method. *Procedia Eng.* **2017**, *184*, 381–389. [CrossRef]
12. Valeri, G.; Koohbor, B.; Kidane, A.; Sutton, M.A. Determining the tensile response of materials at high temperature using DIC and the Virtual Fields Method. *Opt. Lasers Eng.* **2017**, *91*, 53–61. [CrossRef]
13. Shen, X.; Li, M.; Dai, Y.; Zhou, X.; He, P. The effects of preparation temperature on the SiCf/SiC 3D4d woven composite. *Ceram. Int.* **2020**, *46*, 13088–13094. [CrossRef]
14. Wang, C.; Wang, S.; Xie, L.; Liu, Y.; Deng, J.; He, W. Fatigue crack growth behavior of marine steel under variable amplitude loading-combining DIC technique and SEM observation. *Int. J. Fatigue* **2023**, *170*, 107508. [CrossRef]
15. Chen, Y.; Gélébart, L.; Chateau, C.; Bornert, M.; King, A.; Aimedieu, P.; Sauder, C. 3D detection and quantitative characterization of cracks in a ceramic matrix composite tube using X-ray computed tomography. *Exp. Mech.* **2020**, *60*, 409–424. [CrossRef]
16. Zhang, S.; Xue, H.; Wang, S.; Zhang, Y.B. A new method to determine cohesive parameters of elastic-plastic materials based on elastic component of J-integral. *Eng. Fract. Mech.* **2025**, *315*, 110793. [CrossRef]
17. Zhu, J.G.; Xie, H.M.; Li, Y.J.; Hu, Z.X.; Luo, Q.; Gu, C.Z. Interfacial residual stress analysis of thermal spray coatings by miniature ring-core cutting combined with DIC method. *Exp. Mech.* **2014**, *54*, 127–136. [CrossRef]
18. Zhang, H.; Li, X.; Gao, T.; Song, H.; Huang, G. Experimental study on deformation evolution and fracture behaviors of pure titanium at different stress triaxialities. *Eng. Fract. Mech.* **2021**, *258*, 108127. [CrossRef]
19. Wang, D.Q.; Zhu, M.L.; Xuan, F.Z. Correlation of local strain with microstructures around fusion zone of a Cr-Ni-Mo-V steel welded joint. *Mater. Sci. Eng. A* **2017**, *685*, 205–212. [CrossRef]
20. Gao, C.; Iwamoto, T.; Kusaka, T. A DIC-based Taylor impact test by measuring inertia force from acceleration distribution to obtain uniaxial stress–strain behavior of pure aluminum. *Int. J. Impact Eng.* **2024**, *184*, 104834. [CrossRef]
21. Gehri, N.; Mata-Falcón, J.; Kaufmann, W. Automated crack detection and measurement based on digital image correlation. *Constr. Build. Mater.* **2020**, *256*, 119383. [CrossRef]

22. Oka, J.M.; Vaidya, R.U.; Davis, J.T.; Webber, K.W.; Walzel, R.K. Nuclear material container drop testing using finite element analysis with verification using digital Image correlation. *Nucl. Eng. Des.* **2024**, *421*, 113057. [CrossRef]
23. Novich, K.A.; Phero, T.L.; Cole, S.E.; Greseth, C.; Murtrey, M.; Estrada, D.; Jaques, B. Experimental Examination of Additively Manufactured Patterns on Structural Nuclear Materials for Digital Image Correlation Strain Measurements. *Exp. Mech.* **2024**, *64* 1005–1020. [CrossRef]
24. Wang, S.; Xue, H.; Wu, G.; Wang, Z.; Zhao, K.; Ni, C. Investigation of the Inhomogeneous Mechanical and Crack Driving Force of Low Alloy Steel SA508 and Its Welded 309L/308L Stainless Steel Cladding. *Nucl. Sci. Eng.* **2023**, *197*, 623–632. [CrossRef]
25. *GB/T 228.1-2010*; Metallic Materials—Tensile Testing—Part 1: Method of Test at Room Temperature. China Quality and Standards Publishing & Media Company Limited: Beijing, China, 2011.
26. *GB/T 2039-1997*; Metallic Materials—Creep and Stress-Rupture Test in Tension. China Quality and Standards Publishing & Media Company Limited: Beijing, China, 1997.
27. Jordan, B.; Grolleau, V.; Mohr, D. Using surround DIC to extract true stress–strain curve from uniaxial tension experiments. *Int. J Solids Struct.* **2023**, *268*, 112171. [CrossRef]

materials

Article

Extreme Behaviors in Fibrous Material Remodeling: Auxetic to Non-Auxetic Transition and Phase Segregation

Andrea Rodella

Department of Structural and Geotechnical Engineering, Sapienza University of Rome, Via Eudossiana 18, 00185 Rome, Italy; andrea.rodella@uniroma1.it

Abstract: Fibrous materials, prevalent in biological tissues and engineered composites, undergo remodeling in response to mechanical loads, leading to plastic changes in fiber orientation. A previously developed continuum model describes this remodeling process. Building on that framework, the present study examines the extreme behaviors of such materials. Analytical results for the homogeneous response under tensile loading reveal three distinct classes: in class (A), fibers asymptotically approach a specific angle; in class (B), fibers align perpendicularly to the load direction; and in class (C), fibers align either with the load direction or perpendicularly, depending on their initial orientation. Numerical simulations are employed to analyze the non-homogeneous material response in a standard tensile test, demonstrating how differences in behavior arise from the material class and the initial fiber orientation distribution. This investigation focuses on the extreme behaviors of material classes (A) and (C), emphasizing phase segregation and transitions between auxetic and non-auxetic behavior.

Keywords: fibrous material; auxetic; segregation; remodeling; plasticity; variational approach

1. Introduction

Fibrous materials have recently garnered significant interest due to their essential roles in both biological systems and engineered materials. In biological contexts, human tissues and organs, such as skin, cartilage, and connective tissues, are structured with microsized fibers, which are crucial for maintaining mechanical integrity and functionality. Similarly, engineered fibrous materials, including textiles, nonwovens, and composites, are designed with similar architectures and are extensively studied for their unique mechanical properties and diverse applications [1–4].

Fibrous materials, both in biological systems and engineered materials, can undergo plastic reorganization under mechanical loading, leading to changes in fiber orientation. This remodeling process is especially relevant in biological tissues, where mechanical stimuli at the chemo-mechanical level drive the reorganization of collagen fibers. For instance, in the arterial wall, stress-driven collagen fiber remodeling plays a significant role in adapting the tissue to varying loads and pressures [5,6]. The reorientation of fibers in response to stress is a key mechanism for maintaining the tissue structural integrity and functionality, as seen in blood vessels where collagen fibers remodel to accommodate changes in mechanical demands [7,8]. Similarly, in soft tissues such as skin and cartilage, the alignment and reorganization of fibers influence the mechanical properties and adaptability of the tissues, contributing to their ability to withstand dynamic loads [9,10].

These processes are not limited to biological tissues; engineered materials, including hydrogels and fiber-reinforced composites, also exhibit fiber reorientation under mechanical loading. For example, studies have shown that the alignment of fibers in hydrogel-based

 https://doi.org/10.3390/ma18071674

materials can significantly affect their mechanical properties, such as stiffness and elasticity [11,12]. Moreover, understanding these remodeling processes has direct applications in designing materials for soft robotics and tissue engineering, where the ability to control and predict fiber orientation is crucial for optimizing material performance [13,14].

The mechanical properties of these tissues include both elastic [15,16] and inelastic [17–19] responses. Specifically, the extracellular matrix can undergo plastic remodeling, leading to irreversible deformations that significantly affect cell behavior, such as migration and differentiation [19–21].

The effects of reorientation, particularly post-yield deformation, remain largely unexplored in both biological and engineered composites. This study focuses on the plastic threshold mechanism, neglecting viscous effects and other secondary properties. As a result, the asymptotic behavior is determined solely by the choice of free energy, allowing for analytical solutions in two-dimensional (2D) problems [22]. The model is grounded in the principles of continuum mechanics, specifically developed for materials incorporating an internal state variable [23–27]. The model introduced in [22] describes the behavior of fiber reorientation within these materials and demonstrates that remodeling, defined as the reorganization of the internal structure, induces diverse material responses.

In particular, when the analysis is restricted to a 2D beam subjected to traction, the fibers within the material can align themselves in different ways depending on their initial orientation. The fibers may eventually align at a specific angle, perpendicularly to the direction of the applied load, or along the load direction.

Notably, fiber reorganization can induce substantial changes in mechanical behavior, including transitions between auxetic and non-auxetic responses, as well as the emergence of regions with distinct mechanical properties due to phase segregation. While previous studies have established the role of fiber reorientation in influencing material properties, a comprehensive understanding of how these transitions emerge, the conditions under which they occur, and their structural consequences remains incomplete.

The paper addresses this gap by building on the continuum model presented in [22] to investigate the underlying mechanisms driving these phase transitions. By analyzing how fiber realignment governs the macroscopic response of the material, this study provides new insights into the tunability of fibrous materials, with long-term implications for applications in both biological systems and engineered structures.

The paper is organized as follows: Section 2 introduces the continuum model proposed in [22], outlining the key assumptions related to (i) the free energy governing the elastic response and (ii) the plastic reorientation model that describes the dissipation process, defined by the yield function. Section 3 summarizes the main analytical results from [22], focusing on a traction problem in the homogeneous case. Section 4 describes the numerical methods used to obtain the results presented in Section 5, which illustrate phase segregation and the transition from auxetic to non-auxetic behavior. Section 6 discusses how these results can be applied to real-world scenarios. Finally, Section 7 offers concluding remarks and outlines directions for future research.

2. Free Energy and Reversibility Domain Assumptions

Consider a 2D body $\mathcal{B}$ where each point x has a position vector $\mathbf{x} = x_\alpha \mathbf{e}_\alpha$ relative to the origin o. Each point is associated with an oriented fiber, whose direction is given by the unit vector $\mathbf{n}(\vartheta) = \cos\vartheta \mathbf{e}_1 + \sin\vartheta \mathbf{e}_2$, where ϑ is an internal state variable, see Table 1 for the notation. Following the theories [23–25], ϑ influences the free energy, and its evolution dictates dissipation.

Table 1. Notations used in the model.

Symbol	Description
$\mathbf{x}$	Position vector of a point in the body.
ϑ	Internal state variable representing fiber orientation.
$\mathbf{E}$	Linearized strain tensor, based on displacement field $\mathbf{u}$.
Λ	State of the body at time t, consisting of strain $\mathbf{E}$ and fiber orientation ϑ.
$\mathbf{n}$	Unit vector in the fiber direction, depending on the internal state variable ϑ.
$\mathbb{C}$	Elasticity tensor for transversely isotropic material in the fiber direction $\mathbf{n}(\vartheta)$.
μ	Lamé coefficient (shear modulus).
λ	Lamé coefficient (first Lamé constant).
c_1	Material constant describing transverse isotropy.
c_2	Material constant describing transverse isotropy.
E	Young's modulus.
ν	Poisson ratio.
γ	Remodeling torque, associated with changes in fiber orientation ϑ.
$\hat{\eta}$	Dissipation potential, associated with fiber reorientation.
$\bar{\Theta}$	Accumulated fiber rotation over time t.
ψ	Free energy density, a function of $\mathbf{E}$ and ϑ.
δ	Remodeling dissipation.
$\mathcal{E}$	Free energy density, a function of $\mathbf{E}$ and ϑ.
$\mathcal{P}$	Set of valid pairs (c_1, c_2) ensuring positive free energy density.
$\mathcal{S}$	Stiffening set, representing materials where fibers align to enhance stiffness.
$\mathcal{C}$	Compliance set, representing materials where fibers reduce stiffness in their orientation direction.
$\mathcal{R}$	Reversibility domain where the remodeling torque is below a critical threshold.
$\partial\mathcal{R}$	Boundary of the reversibility domain: yield set.

The state of $\mathcal{B}$ at time t is defined by $\Lambda = \{\mathbf{E}, \vartheta\}$, where $\mathbf{E} = \text{sym}\nabla\mathbf{u}$ is the linearized strain based on the displacement field $\mathbf{u}$. The free energy density is a quadratic function of the strain:

$$\psi = \widehat{\psi}(\mathbf{E}, \vartheta) = \frac{1}{2}\mathbb{C}(\mathbf{n}(\vartheta))\mathbf{E} \cdot \mathbf{E}. \tag{1}$$

Here, $\mathbb{C}(\mathbf{n}(\vartheta))$ is the elasticity tensor for a transversely isotropic material with respect to $\mathbf{n}(\vartheta)$. In 2D, the free energy density is as follows:

$$\widehat{\psi}(\mathbf{E}, \vartheta) = \mu\|\mathbf{E}\|^2 + \frac{\lambda}{2}(\text{tr}\mathbf{E})^2 + c_1(\text{tr}\mathbf{E})\,\mathbf{E}\,\mathbf{n}(\vartheta)\cdot\mathbf{n}(\vartheta) + c_2\left(\mathbf{E}\,\mathbf{n}(\vartheta)\cdot\mathbf{n}(\vartheta)\right)^2, \tag{2}$$

where λ and μ are the Lamé coefficients, and c_1 and c_2 characterize the transverse isotropy. With fixed Lamé coefficients ($\mu > 0$, $2\mu + \lambda > 0$), the pair (c_1, c_2) must belong to a set $\mathcal{P}$ to ensure the positiveness of the energy density:

$$\mathcal{P} = \left\{ (c_1, c_2) \in \mathbb{R} : \; c_2 > \frac{c_1^2 - 4\mu c_1 - 4\mu(\mu+\lambda)}{2(2\mu+\lambda)} \right\}. \tag{3}$$

This set is represented by the red dashed parabola plotted in Figure 1. To understand the role of c_1 and c_2, the author recall the Young modulus and Poisson ratio definitions for an uniaxial traction test $\bar{\mathbf{T}} = \sigma\mathbf{t}\otimes\mathbf{t}$ in the direction $\mathbf{t} = \cos\alpha\mathbf{e}_1 + \sin\alpha\mathbf{e}_2$. The corresponding strain is $\bar{\mathbf{E}} = \mathbb{C}^{-1}(\mathbf{n}(\vartheta))\bar{\mathbf{T}}$. Thus, Young modulus and Poisson ratio depend on the angle $(\alpha - \vartheta)$ between the testing direction and the fiber orientation:

$$E(\alpha, \vartheta) = \hat{E}(\alpha, \vartheta) := \frac{\bar{\mathbf{T}}\mathbf{t}\cdot\mathbf{t}}{\bar{\mathbf{E}}\mathbf{t}\cdot\mathbf{t}}; \qquad \nu(\alpha, \vartheta) = \hat{\nu}(\alpha, \vartheta) := \frac{\bar{\mathbf{E}}\mathbf{t}^{\perp}\cdot\mathbf{t}^{\perp}}{\bar{\mathbf{E}}\mathbf{t}\cdot\mathbf{t}}. \tag{4}$$

Polar plots illustrate how Young modulus varies with $(\alpha - \vartheta)$, as shown in the insert of Figure 1a. The stiffening set $\mathcal{S}$ is defined by ensuring that the ratio of Young moduli parallel and perpendicular to the fiber direction ($E_{\parallel} = \hat{E}(\alpha = \vartheta, \vartheta)$ and $E_{\perp} = \hat{E}(\alpha = \vartheta - \pi/2, \vartheta)$) is greater than 1 as follows:

$$\frac{E_{\parallel}}{E_{\perp}} = 1 + \frac{2(c_1 + c_2)}{2\mu + \lambda} > 1 \quad \Longrightarrow \quad \mathcal{S} = \{(c_1, c_2) \in \mathcal{P} : \ c_1 + c_2 > 0\}. \tag{5}$$

This set is represented by the yellow area in Figure 1a. Conversely, the compliance set $\mathcal{C} = \mathcal{P} \setminus \mathcal{S}$ represents materials where fibers reduce stiffness in their orientation direction. This set is depicted in gray in Figure 1a. For more details on the possible application of this set, the reader is referred to [28,29], where the authors discuss the potential use of these materials in modeling the tethering of fibers in the extracellular matrix of biological tissues.

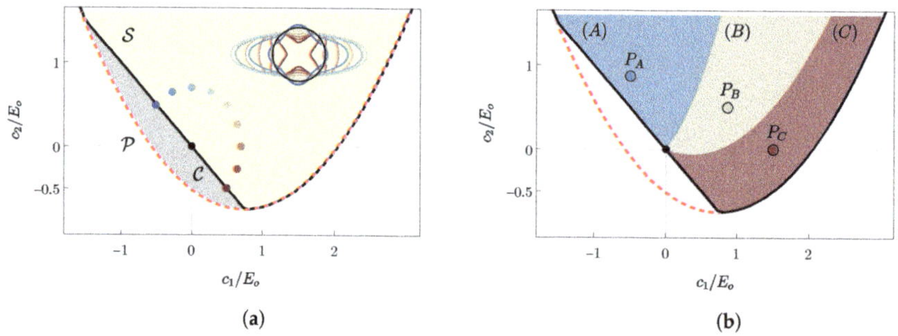

Figure 1. **(a)** The red dashed line confines the set where $\{c_1, c_2\}/E_0$ ensures positive energy (Equation (3)), with E_0 being the isotropic Young modulus. The yellow area, a subset of $\mathcal{P}$, represents the stiffening set (Equation (5)). The insert shows polar plots of Young modulus $\hat{E}(\alpha, \vartheta = 0)$ for $\alpha \in [0, 2\pi]$, corresponding to colored dots from the stiffening set $\mathcal{S}$. **(b)** Colored areas depict asymptotic behaviors of fiber orientation ϑ in traction tests (see [22]): alignment at a given angle (blue), perpendicular alignment to the load (gray), and alignment either perpendicularly or with the load, depending on initial orientation (red).

Within the Generalized Standard Materials framework [25], dissipation is described by the dissipation potential, a convex and positive function of the internal state variable rates. Since $\mathbf{n} \cdot \mathbf{n} = 1$ and $\dot{\mathbf{n}} = \beta \mathbf{n}^{\perp}$ with $\mathbf{n}^{\perp} := \partial_{\vartheta}\mathbf{n}(\vartheta) = -\sin\vartheta \mathbf{e}_1 + \cos\vartheta \mathbf{e}_2$, then the scalar product between $\dot{\mathbf{n}} \cdot \mathbf{n} = 0$. Thus, from Equation (2), the energy release rate takes the form:

$$\begin{aligned} -\partial_{\vartheta}\hat{\psi}(\mathbf{E}, \vartheta)\dot{\vartheta} &= -\partial_{\mathbf{n}}\hat{\psi}(\mathbf{E}, \mathbf{n}(\vartheta)) \cdot \dot{\mathbf{n}}(\vartheta) \\ &= -2\Big(c_1 \mathrm{tr}\mathbf{E} + 2c_2 \mathbf{E}\mathbf{n}(\vartheta) \cdot \mathbf{n}(\vartheta)\Big) \mathbf{E}\mathbf{n}(\vartheta) \cdot \beta \mathbf{n}^{\perp}(\vartheta) \\ &= -\beta\gamma(\mathbf{E}, \vartheta), \end{aligned} \tag{6}$$

where $\gamma(\mathbf{E}, \vartheta)$ is the remodeling torque associated with changes in fiber orientation ϑ. To satisfy the fundamental inequality, [25], the energy release rate at $\mathbf{E}$ must exceed that at any admissible $\tilde{\mathbf{E}}$:

$$-\beta[\gamma(\mathbf{E}, \vartheta) - \gamma(\tilde{\mathbf{E}}, \vartheta)] > 0 \tag{7}$$

for any $\mathbf{E}$ and $\tilde{\mathbf{E}}$ in the reversibility domain, $\mathcal{R}(\eta)$. This domain comprises symmetric strain tensors for which the remodeling torque remains below a critical threshold $\eta > 0$:

$$\mathcal{R}(\eta) = \left\{ \mathbf{E} \in \mathrm{Sym} : \sup_{\vartheta \in [-\pi/2, +\pi/2]} |\gamma(\mathbf{E}, \vartheta)| \leq \eta \right\}. \tag{8}$$

This set represents strains for which the fiber orientation ϑ remains constant, resulting in purely elastic material behavior. Its boundary, named *yield set*, is denoted by $\partial\mathcal{R}(\eta) = \{\mathbf{E} : \sup_\vartheta |\gamma(\mathbf{E},\vartheta)| = \eta\}$. Following [25,30], the dissipation rate is given by the Legendre transform:

$$d(\vartheta)(\dot{\vartheta}) = \sup_{\mathbf{E}\in\mathcal{R}(\eta)} \Big(-\partial_\vartheta \hat{\psi}(\mathbf{E},\vartheta)\dot{\vartheta}\Big) = \eta\,\beta\,|\dot{\vartheta}| = \bar{\eta}|\dot{\vartheta}|. \tag{9}$$

Considering a quasi-static process over a time interval t, the total dissipation is

$$\delta = \hat{\delta}(t) = \int_0^t d(\vartheta(\tau))(\dot{\vartheta}(\tau))\,d\tau = \bar{\eta} \int_0^t |\dot{\vartheta}(\tau)|\,d\tau =: \bar{\eta}\,\bar{\Theta}(t), \tag{10}$$

where $\bar{\Theta}(t)$ is the accumulated fiber rotation over time t. Finally, the total energy is the integral of the free energy density, Equation (1), plus the remodeling dissipation, Equation (10), over the body $\mathcal{B}$:

$$\mathcal{E}\big(\mathbf{E}(t),\vartheta(t)\big) = \int_{\mathcal{B}} \Big(\hat{\psi}(\mathbf{E}(t),\vartheta(t)) + \hat{\delta}(t)\Big)\,dA. \tag{11}$$

3. Analytical Solution for the Homogeneous Material in a Traction Problem

Consider the traction problem sketched in Figure 2, where a rectangular sample of length L and height H is free on the upper and lower edges and allowed to slide on the left side. The points on the right side experience a horizontal displacement of $\bar{\mathbf{u}} = \varepsilon L\,\mathbf{e}_1$, whilst their vertical displacement remains unconstrained.

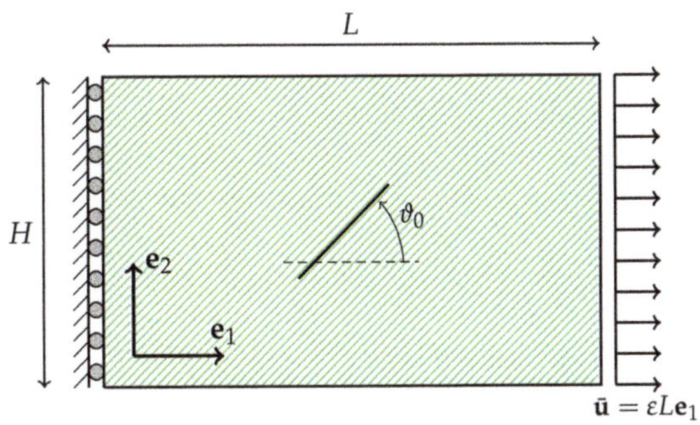

Figure 2. Sketch of the setup for the traction problem, where ϑ_0 represents the initial orientation of the fibers.

In [22], this problem is presented a complete analytical characterization of the homogeneous case in terms of strain and fiber rotation, as the imposed horizontal displacement $\bar{\mathbf{u}} = \varepsilon\,L\mathbf{e}_1$ monotonically increases starting from $\varepsilon = 0$. The fiber orientation ϑ evolves according to the following Karush Kuhn Tacker (KKT) conditions [31]:

$$\dot{\vartheta} \geq 0, \qquad |\gamma(\mathbf{E},\vartheta)| - \eta \leq 0, \qquad \dot{\vartheta}(|\gamma(\mathbf{E},\vartheta)| - \eta) = 0. \tag{12}$$

Hence, it is possible to find an asymptotic value of the fiber orientation for $\varepsilon \to \infty$ once the material reaches the plastic regime:

$$\vartheta_\infty(\lambda,\mu,c_1,c_2) = \frac{1}{2}\arccos\left(\frac{2\mu(c_1+c_2)}{c_1^2 - 2c_2(\mu+\lambda)}\right). \tag{13}$$

In summary, the stiffening materials are identified in three classes as represented in different colored regions in Figure 1b:

Class (A): Materials in class (A) exhibit a minimum of the free energy ψ at $0 < \vartheta_\infty < \pi/2$ (it suffices to consider $\vartheta \in [0, \pi/2]$ as $\vartheta \equiv \vartheta + \pi$ and, for the symmetry of the problem, $\vartheta \equiv -\vartheta$) and two local maxima at $\vartheta = \{0, \pi/2\}$. See, for instance, blue dots to the top left of Figure 3 for an illustration. The evolution of the homogeneous traction problem, starting from an initial uniform fiber orientation $\vartheta_0 \in [0, \pi/2]$, is depicted in the bottom left of Figure 3. The fiber orientation evolves with jumps if ϑ_0 lies on an unstable branch (dashed blue lines in Figure 3 bottom left, corresponding to $0 \leq \vartheta_0 < \vartheta^A_{s1}$ or $\vartheta^A_{s2} < \vartheta_0 \leq \pi/2$), or continuously if ϑ_0 is initially on a stable branch (continuous blue lines in Figure 3 bottom left, corresponding to $\vartheta^A_{s1} \leq \vartheta_0 < \vartheta_\infty$ or $\vartheta_\infty < \vartheta_0 \leq \vartheta^A_{s2}$). From a physical standpoint, materials in class (A) exhibit preferred fiber orientations that do not align with principal loading directions, leading to anisotropic stiffening effects that are particularly useful in biological tissues or engineered composites where directional reinforcement is required. For example, in muscle fibers or fibrous composite laminates, such energy landscapes help in tailoring mechanical responses to external forces while maintaining structural integrity [32,33].

Class (B): Materials in class (B) exhibit a single minimum of ψ at $\vartheta = \pi/2$, with $\vartheta = 0$ representing a maximum. Furthermore, ϑ_∞ does not exist within the real numbers. Refer to the middle top of Figure 3 for the energy landscape and the middle bottom for the evolutionary paths of a generic initial fiber orientation, illustrated by continuous gray lines with arrows. Physically, class (B) materials favor alignment of fibers in a single dominant direction ($\vartheta = \pi/2$), which suggests a material design where unidirectional reinforcement is beneficial. Such a configuration is particularly relevant in fiber-reinforced polymers and natural materials such as tendons, where maximizing stiffness along a specific axis is crucial for load-bearing efficiency and mechanical performance under tension.

Class (C): Materials in class (C) exhibit two minima of ψ at $\vartheta = \{0, \pi/2\}$, with ϑ_∞ corresponding to a maximum. Refer to the top right panel of Figure 3 for the energy landscape and the bottom right for the evolutionary paths of a generic initial fiber orientation, illustrated by continuous gray lines with arrows. As indicated by the large red arrows, ϑ_∞ is a critical point, meaning that a fiber initially oriented at $\vartheta_\infty + \alpha$, with α being a small perturbation, will reorient towards $\pi/2$. Conversely, if the initial orientation is $\vartheta_\infty - \alpha$, the final orientation will be 0. This scenario allows for the observation of phase segregation in a material with particular distributions of fiber orientation. In practical terms, class (C) materials exhibit bistable behavior, which is crucial for applications where controlled anisotropy and phase separation are desirable. This includes smart materials, such as shape-memory polymers or adaptive bio-inspired structures, where switching between two stable configurations under mechanical stimuli can be leveraged for tunable stiffness, actuation, or self-healing capabilities. Additionally, phase segregation effects in this class are particularly relevant for tissue engineering, where fiber alignment can dictate cellular growth patterns and mechanical functionality in biomimetic scaffolds [34,35].

By classifying materials based on these energy profiles, this framework provides a fundamental understanding of how the fiber orientation process influences mechanical behavior. This insight aids in designing materials with tailored anisotropic properties for various engineering and biomedical applications, ranging from high-performance composites to biomimetic tissues.

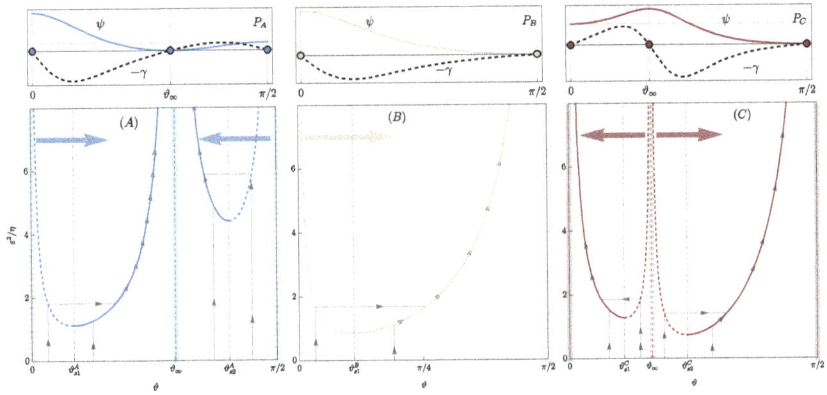

Figure 3. The top row illustrates the free energy landscape ψ (solid colored lines) and its derivative $\partial_\vartheta\psi = -\gamma$ (dashed black line) for material classes (A), (B), and (C), corresponding to points P_A, P_B, and P_C in Figure 1b, distinguished by blue, white, and red colors, respectively. The bottom row depicts the evolution, in the yield set, of the fiber orientation ϑ from various initial orientations; continuous gray lines indicate the evolutionary paths, which can be continuous or exhibit jumps.

4. Numerical Implementation

The energetic properties of the problem solution and the general conditions offered by the incremental energy minimization framework [36] suggest the alternating minimization (AM) strategy as a natural resolution algorithm. In this specific case, the AM consists of two sequential minimization steps repeated over each time increment until convergence: first, the total energy is minimized with respect to the displacement field, $\mathbf{u}$, while keeping the fiber rotational field, ϑ, fixed. Second, the total energy is minimized with respect to ϑ, while keeping $\mathbf{u}$ fixed. This process is detailed in the pseudo-code Algorithm A1.

The first step is equivalent to solving a linear elasticity problem. The second step is a nonlinear problem, but it is local in space, allowing for minimization via a local projection method, such as a modified Newton-Raphson scheme. The problem requires discretization in both time and space. The time discretization is merely a path-ordering parameter since no dynamics is involved.

Space Discretization: The domain $\mathcal{B} = \{(x_1, x_2) : x_1 \in (0, L) \wedge x_2 \in (0, H)\}$ is discretized using an unstructured mesh. The displacement field $\mathbf{u}$ is approximated using a piece-wise affine finite element space [37] over $\mathcal{B}$. The fiber rotational field ϑ, due to its intrinsic locality (a feature common in rate-independent plasticity problems), is projected onto a discontinuous finite element space over $\mathcal{B}$.

Time Discretization: The time interval $[0, T]$ is discretized in n non-uniform time steps dt_i. The accumulated fiber rotation is discretized as follows:

$$\bar{\Theta}(t_i) \simeq \bar{\Theta}_i = \bar{\Theta}_{i-1} + \|\vartheta_i + \vartheta_{i-1}\|. \tag{14}$$

At each time step i, the solution, representing the state Λ of the sample, must be a local minimum of the total energy with respect to both fields:

$$\{\mathbf{u}_i, \vartheta_i; \bar{\Theta}_i\} = \underset{\mathbf{u},\vartheta}{\operatorname{argmin}}\{\mathcal{E}_i(\mathbf{u}, \vartheta; \bar{\Theta})\}, \qquad \mathcal{E}_i(\mathbf{u}, \vartheta; \bar{\Theta}) = \int_B \psi_i(\mathbf{u}, \vartheta) + \delta_i(\vartheta; \bar{\Theta}), \tag{15}$$

subject to the Dirichlet boundary conditions $\mathbf{u}(0, x_2) \cdot \mathbf{e}_1 = 0$ and $\mathbf{u}(L, x_2) \cdot \mathbf{e}_1 = \bar{u}(\varepsilon_i)$, where $\mathcal{E}_i$ represents the time-discretized form of the total energy in Equation (11).

The code implementation uses `python` as an interface to FEniCS (https://fenicsproject.org, accessed on 5 April 2025) [38,39], an open-source computing platform for solving partial differential equations (PDEs).

5. Numerical Results

This section presentes the numerical results obtained by solving the problem described in Section 2 using the numerical implementation detailed in Section 4. Starting from the analytical solution discussed in Section 3, this study explores two different scenarios: (i) the transformation from an auxetic to a non-auxetic mechanical behavior of a material in class (A) under specific initial fiber orientation conditions, and (ii) the segregation process of the fiber orientation field which is characteristic of materials in class (C). Materials in class (B) are not considered in this work, as they do not exhibit any extreme behavior. Once perturbed, fibers in class (B) materials tend to reorient monotonically toward a single stable configuration, making their evolution relatively straightforward compared to the bistable or metastable behaviors of classes (A) and (C). However, such materials are particularly relevant in applications where a predictable and stable anisotropic response is desired, such as tendon-like biological tissues or fiber-reinforced composites designed for unidirectional strength. For further details about this material, the reader is referred to [22].

5.1. Auxetic to Non-Auxetic Trasformation

The present section focuses on a specific material in class (A), represented by the blue dot, P_A, in Figures 1b and 4. This material has anisotropic parameters $c_1 = -0.75$ and $c_2 = 1.3$, while the Lamé coefficients are chosen as $\lambda = \mu = 3/8$. Equation (13) provides the asymptotic orientation, which is $\vartheta_\infty \simeq 54°$. As seen in Section 3, ϑ_∞ is a minimum of the free energy and coincides with the limit orientation that the fibers will tend towards during the loading process, regardless of the initial orientation.

In Figure 4, two similar schematics are shown where the black region represents the set of anisotropic coefficients that exhibit auxetic behavior at ϑ_∞, denoted as $\mathcal{A}_\infty = \{(c_1, c_2) \in \mathcal{S} : \nu(\vartheta_\infty) < 0\}$. The material under considerations, represented by point P_A, does not belong to this area, indicating that the asymptotic orientation does not exhibit auxetic behavior. Conversely, the red region represents the auxetic set $\mathcal{A}_0 = \{(c_1, c_2) \in \mathcal{S} : \nu(\vartheta_0) < 0\}$. In panel (a), $\vartheta_0 = 6° < \vartheta^A_{s1}$, while in panel (b), $\vartheta_0 = \vartheta^A_{s1} \simeq 15°$. Although the black regions are identical in both panels, the red areas differ. On the left, P_A belongs to $\mathcal{A}_0$, while on the right, it does not. This indicates that the material P_A is a good candidate to perform the transition from auxetic to non-auxetic: with an initial orientation $\vartheta_0 = 6°$, it behaves like an auxetic material, but when the fiber reaches its limit orientation, the material no longer exhibits auxetic behavior.

In Figure 5a, the reorientation path is represented by red lines with arrows as ε increases, while Figure 5b shows the complete landscape of the Poisson ratio, given by Equation (4). Figure 6a–d presents four snapshots taken along this loading path: the snapshot (a) shows the initial orientation and the undeformed shape of the sample. As ε increases, the homogeneous material reaches the boundary of the reversibility domain, labeled S_0 in Figure 5. The behavior is auxetic, as seen in the deformed shape in Figure 6b snapshot S_0, where, for instance, the left side of the deformed shape is larger than the same side of the original undeformed shape (black rectangle). Upon reaching this limit, the orientation evolves, with ϑ jumping from $\vartheta_0 = 6°$ to 30°, labeled S_1 in Figure 6c. From this snapshot and the Poisson ratio landscape, one can observe that the behavior is no longer auxetic until it reaches the asymptotic value of 54°, Figure 6d.

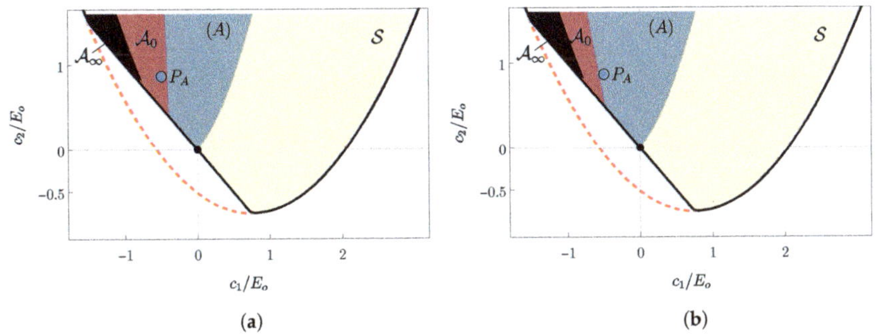

Figure 4. The black region represents the set of anisotropic coefficients that exhibit auxetic behavior at the asymptotic orientation ϑ_∞, denoted as $\mathcal{A}_\infty$. The red region represents the auxetic set $\mathcal{A}_0$. The panel (**a**) shows the latter set with $\vartheta_0 = 6°$, while the panel (**b**) shows the same set with $\vartheta_0 = \vartheta_{s1}^A \simeq 15°$.

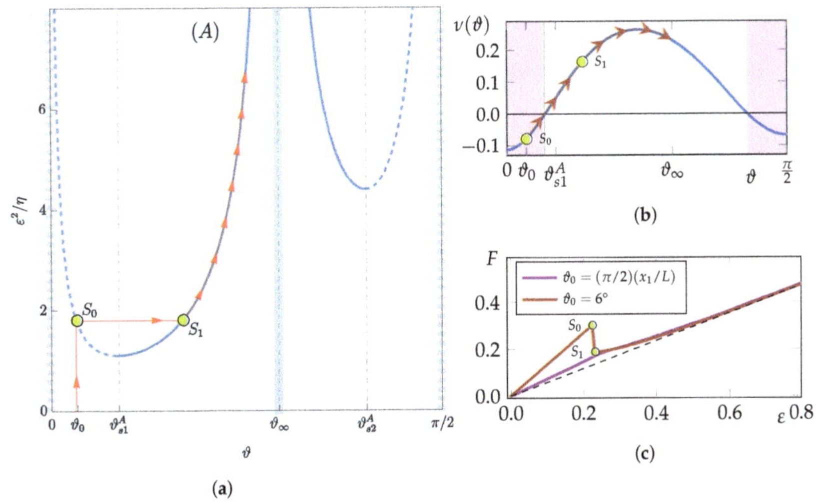

Figure 5. (**a**) The reorientation path in the plane $\{\vartheta, \varepsilon^2/\eta\}$ is represented by the continuous red line with arrows. (**b**) Poisson ratio landscape for the material in class (A) with $c_1 = -0.75$ and $c_2 = 1.3$. (**c**) Force F as a function of the applied displacement ε.

The same process is analyzed for a material in class (A) with a linear distribution of the initial fiber orientation, $\vartheta_0 = (\pi/2)(x_1/L)$, as shown in the right panel of Figure 6e–h. At small load levels ε, the left and right extremities of the deformed shape exhibit auxetic behavior, characterized by an increase in size, as indicated by the violet region in the Poisson ratio landscape. As the loading progresses, the material transitions towards the asymptotic orientation $\vartheta_\infty \simeq 54°$, which is nearly achieved everywhere in the configuration shown in Figure 6h. However, it is important to note that the exact asymptotic value is only fully reached as $\varepsilon \to \infty$, as depicted in Figure 5a. Additionally, the yield set for fibers oriented at $\vartheta > \vartheta_\infty$ is reached at higher values of ε compared to fibers oriented at $\vartheta < \vartheta_\infty$.

Finally, in Figure 5c, the material response is shown in terms of force F as a function of the applied displacement ε. The force is computed at each load step as the integral of the traction vector $\mathbf{T}\mathbf{n}_\Gamma$ over the boundary, $\partial_u\mathcal{B}$, where the displacement is applied:

$$F = \int_{\partial_u\mathcal{B}} \mathbf{T}\mathbf{n}_\Gamma \, d\Gamma, \tag{16}$$

where $\mathbf{n}_\Gamma$ is the normal unit vector to that boundary. In this panel, the red line represents the response of the material with initial orientation $\vartheta_0 = 6°$, with the drop S_0–S_1 indicating the sudden reorientation. The purple line represents the response of the material with a linear distribution of the initial orientation $\vartheta_0 = (\pi/2)(x_1/L)$. Both responses tend towards the same asymptotic behavior at large ε, represented by the dashed black line.

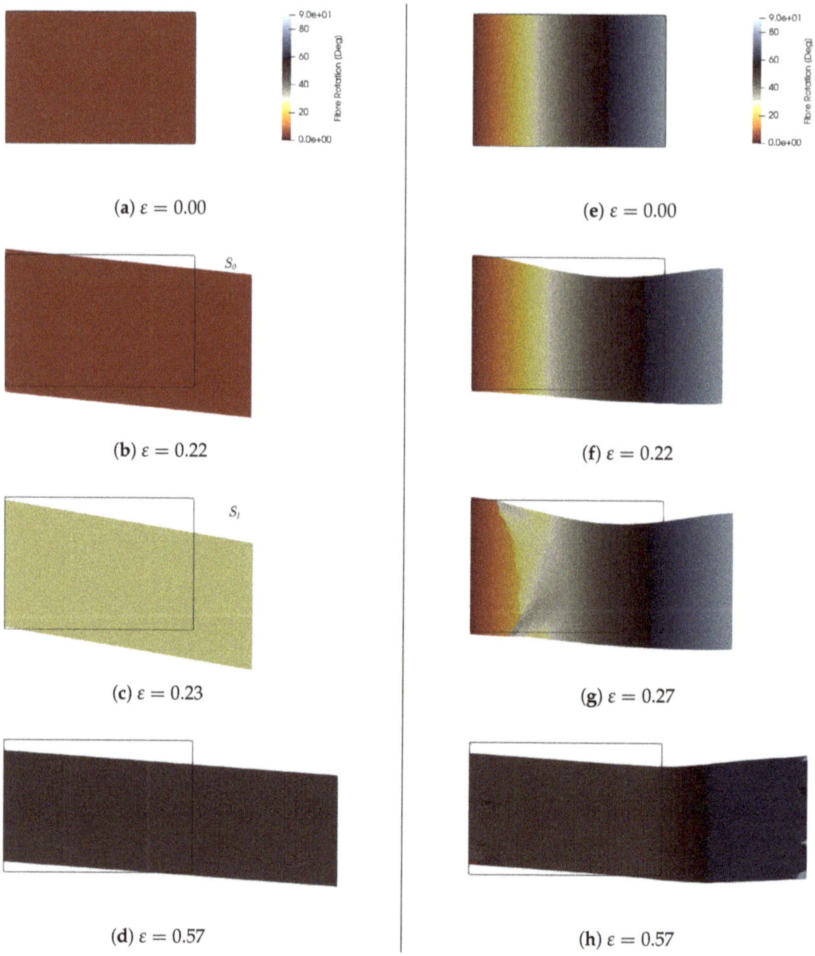

Figure 6. Auxetic to non-auxetic transition. Material in class (A) with $c_1 = -0.75$ and $c_2 = 1.3$. (**a**–**d**) Left panel from top to bottom: (**a**) initial undeformed configuration with constant distribution of $\vartheta_0 = 6°$ followed by (**b**) the auxetic behavior, (**c**) the transition to non-auxetic, and finally (**d**) the asymptotic value $\vartheta_\infty \simeq 54°$. (**e**–**h**) Right panel from top to bottom: (**e**) initial undeformed configuration with linear distribution of $\vartheta_0 = \pi/2x_1/L$ followed by (**f**) the simultaneous auxetic (sides) and non-auxetic (middle) behaviors , (**g**) the propagation of the non-auxetic phase, and finally (**h**) reaching the limit value almost everywhere $\vartheta_\infty \simeq 54°$.

5.2. Segregation Process

To illustrate the segregation process, the analysis focuses on a material from class (C), represented by the red dot P_C in Figure 1b. This material has anisotropic parameters $c_1 = 1.5$ and $c_2 = 0$, while the Lamé coefficients are chosen as $\lambda = \mu = 3/8$. The asymptotic orientation is $\vartheta_\infty \simeq 30°$. For this specific material class (C), ϑ_∞ becomes the maximum of the free energy and corresponds to the critical point, which means that a fiber initially oriented as $\vartheta_\infty \pm \alpha$, with α a small perturbation, will reorient towards $\pi/2$ or 0, respectively. This section presents two study cases, in which the author explores the evolutions of different non-uniform distributions of initial orientation $\vartheta_0(\mathbf{x})$. The first one has a piece-wise linear distribution along the coordinate x_1:

$$\vartheta_0^i(x_1) = \begin{cases} \frac{\pi}{5L}x_1 & \text{if } 0 \le x_1 \le \frac{L}{2}, \\ \frac{\pi}{2L}x_1 & \text{if } \frac{L}{2} > x_1 > L. \end{cases} \tag{17}$$

This is also plotted in red in the left panel of Figure 7 along the sample diagonal, and is shown by the white line in Figure 8a.

The second study case has an initial distribution that follows the function:

$$\vartheta_0^{ii}(\mathbf{x}) = \frac{\pi}{2L}x_1 \cos\left(\frac{\pi}{2H}x_2\right) \quad \text{with } x_1 \in [0, L] \text{ and } x_2 \in [0, H], \tag{18}$$

this becomes 0 along the left and top edges and reaches $\pi/2$ in the bottom-right corner $x_1 = L$ and $x_2 = 0$; see the contour plot of snapshot on the panel (e) in Figure 8 and blue line in the left panel of Figure 7 for the effective distribution along the diagonal of the sample.

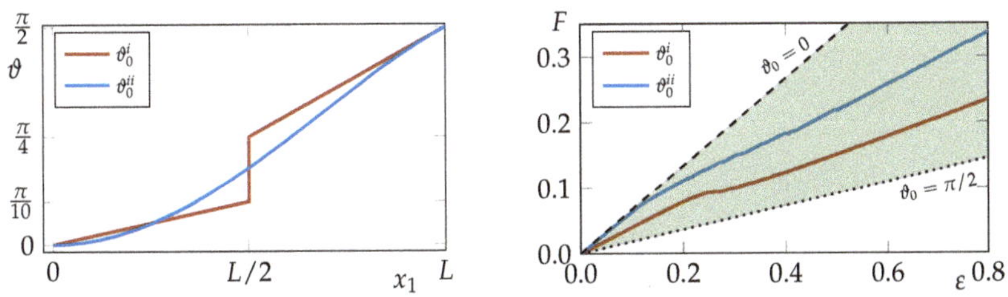

Figure 7. (Left) Initial distributions of the fiber orientation for cases: (i) ϑ_0^i Equation (17) and (ii) ϑ_0^{ii} Equation (18) both plotted along the dashed white line in Figure 8 top row. (Right) Force F as a function of the applied displacement ε.

The evolution of the fiber orientation for these two cases is depicted in Figure 8. The left panel shows the deformed shape and the evolution, as a contour plot, of the fiber orientation field for the piece-wise linear distribution given by Equation (17), ϑ_0^i. The right panel shows the deformation and reorientation for ϑ_0^{ii}, given by Equation (18). The top row of Figure 8 shows the initial distribution of the fiber orientation, while the bottom row shows the final distribution. In both cases, the segregation process is evident, particularly in the second study case where one can observe the propagating front of segregation. The final distributions show fibers reaching the two minima at 0° and 90°, polarizing the sample in two regions.

Finally, the force–displacement curve is shown in the right panel of Figure 7. The red line represents the response of the material with the piece-wise linear distribution of the initial orientation, while the purple line represents the response of the material with the distribution (18). Both responses are confined within the green cone, which is bounded by the responses of the material characterized by initial fiber orientations of 0° (dashed black line) and 90° (dotted black line).

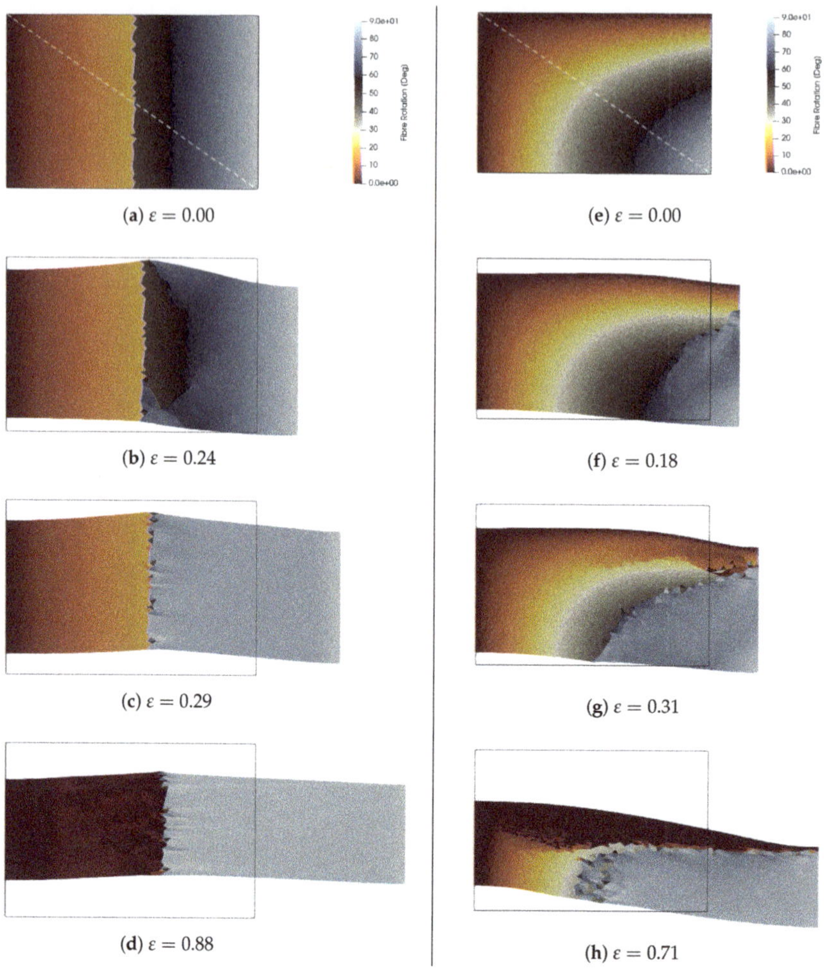

Figure 8. Segregation process. Material in class (C) with $c_1 = 1.5$ and $c_2 = 0$. (**a**–**d**) Left panel from top to bottom: (**a**) initial undeformed configuration with piece-wise linear distribution of ϑ_0^i given by Equation (17), (**b**) the segregation front on the right half of the sample, (**c**) the completely segregated half sample, and (**d**) the final distribution: the red part is at 0° while the grey half has an orientation equal to 90°. (**e**–**h**) Right panel from top to bottom: (**e**) initial undeformed configuration with distribution of ϑ_0^{ii} given by Equation (18), (**f**) the segregation process is activated from the bottom right corner, in (**g**) the segregation front starts to appear, and (**h**) reports the final segregated distribution.

6. Practical Implications

One key aspect of the results is the emergence of phase segregation, where distinct fiber orientations develop within different regions of the material. This phenomenon leads to localized variations in mechanical properties, potentially influencing failure mechanisms, toughness, and adaptability. In engineering applications, such behavior can be leveraged to design materials with tailored mechanical responses, such as controlled deformation patterns or enhanced energy dissipation. In biological tissues, phase segregation may play a role in structural adaptation processes, allowing different regions to develop specialized mechanical functions over time [40,41].

Furthermore, the transition from auxetic to non-auxetic behavior, governed by fiber alignment, offers opportunities for designing materials with tunable mechanical properties. Auxetic materials, characterized by their negative Poisson ratio, are known for their enhanced fracture resistance and superior shock absorption [42,43]. A deeper understanding of this transition can facilitate the development of smart materials for applications in biomedical implants, soft robotics, and impact-resistant structures.

By capturing these extreme behaviors, this study lays the foundation for designing advanced fibrous materials with programmable mechanical properties. The findings contribute to optimizing composite structures, improving the durability of biomimetic materials, and refining computational models for predicting material responses under various loading conditions. Moreover, the theoretical framework and numerical simulations presented here provide a basis for future experimental validation and practical implementation in real-world applications.

7. Concluding Remarks and Future Perspectives

This work presents a comprehensive continuum model for fibrous materials undergoing plastic remodeling, leading to significant changes in fiber orientation and, consequently, their mechanical response. The study demonstrates that such remodeling processes can result in complex material behaviors, including phase segregation and transitions between auxetic and non-auxetic properties, with far-reaching implications for both engineered materials and biological tissues.

Through a combination of analytical and numerical approaches, the author demonstrates how the reorientation of fibers can drastically alter the mechanical response of the material. Specifically, three distinct classes of material behavior were identified based on the peculiarities of the free energy function: materials that exhibit a stable fiber orientation at an intermediate angle, materials that align fibers perpendicularly to the load direction, and materials that can exhibit either behavior depending on the initial fiber orientation. This classification underscores the crucial role of fiber orientation in governing the mechanical properties of fibrous materials.

Beyond theoretical insights, the findings have practical relevance for material design and optimization. The emergence of phase segregation—where different regions of the material develop distinct fiber orientations—has implications for localized mechanical properties, influencing structural toughness, failure mechanisms, and adaptability. This phenomenon could be exploited to engineer materials with controlled deformation patterns, improved energy dissipation, or enhanced durability in applications ranging from soft robotics to impact-resistant structures. Similarly, the ability to transition between auxetic and non-auxetic behaviors offers exciting possibilities for tunable materials. Auxetic structures, known for their enhanced fracture resistance and shock absorption, could be tailored for biomedical implants, protective equipment, and morphing materials capable of adjusting their mechanical response in real time.

Numerical simulations have provided further insights into the evolution of fiber orientation under applied loads. These insights pave the way for the development of fibrous materials with programmable mechanical properties, optimizing composite structures and biomimetic materials for specialized applications.

Future work will address the current limitations by extending the model to three-dimensional settings, enabling a more comprehensive understanding of remodeling in real-world scenarios. Additionally, incorporating viscous effects will enhance the model's ability to capture time-dependent material responses, improving its predictive accuracy and facilitating closer alignment with experimental observations.

Overall, this study lays a solid foundation for advancing the understanding of fibrous materials undergoing plastic remodeling, offering new opportunities for the design of materials with tailored mechanical properties for engineering, biomedical, and industrial applications.

Funding: A.R. acknowledges the financial support from the Italian Ministry of University and Research (MUR) under the Project PRIN 2022: New eco-friendly building materials inspired by ancient constructions (Prot. 2022Y2RHHT).

Data Availability Statement: No data were used to support this study.

Acknowledgments: A.R. acknowledges the Italian INdAM-GNFM.

Conflicts of Interest: The authors declare no conflicts of interest.

Appendix A

In this appendix, the author reports the algorithm presented in Section 4, which outlines the alternating minimization scheme used to solve the coupled problem of displacement and fiber remodeling in fibrous materials.

Algorithm A1 Alternating Minimization Scheme

Require: Initial values $\{\mathbf{u}_0, \vartheta_0, \bar{\Theta}_0\}$ and Dirichlet boundary conditions
Ensure: Solution fields $\{\mathbf{u}_n, \vartheta_n, \bar{\Theta}_n\}$ at the final time step

Initialize: $i = 0$, $\mathbf{u}_0 = \mathbf{0}$, $\vartheta_0 = 0$, $\bar{\Theta}_0 = 0$
for $i = 1, 2, \ldots, n$ **do**
  Update Dirichlet boundary conditions based on the current time step
  while $\mathbf{u}_{err} > \mathbf{u}_{tol}$ AND $\vartheta_{err} > \vartheta_{tol}$ **do**
    Solve the linear elastic problem for $\mathbf{u}_i$:
    $\mathbf{u}_i = \arg\min_{\mathbf{u}} \{\mathcal{E}_i(\mathbf{u}, \vartheta_{i-1}; \bar{\Theta}_{i-1})\}$
    Evaluate the yield function $\gamma(\mathbf{u}_i, \vartheta_{i-1})$
    if $\gamma(\mathbf{u}_i, \vartheta_{i-1}) \leq \bar{\eta}$ **then**
      Remodeling does not occur; $\vartheta_i = \vartheta_{i-1}$
    else
      Initialize local projection iteration: $j = 0$, $\vartheta_i^0 = \vartheta_{i-1}$
      while $\vartheta_{err,proj} > \vartheta_{tol,proj}$ **do**
        Increment projection iteration: $j = j + 1$
        Solve the remodeling problem via local projection:
        $\vartheta_i^j = \text{Projection}(\vartheta_i^{j-1} + \Delta\vartheta^j)$
        Update projection error: $\vartheta_{err,proj} = \|\vartheta_i^j - \vartheta_i^{j-1}\|_{L_2}$
      end while
      $\vartheta_i = \vartheta_i^j$
    end if
    Update displacement error: $\mathbf{u}_{err} = \|\mathbf{u}_i - \mathbf{u}_{i-1}\|_{H_1}$
    Update fiber rotation error: $\vartheta_{err} = \|\vartheta_i - \vartheta_{i-1}\|_{L_2}$
  end while
  Update accumulated fiber rotation: $\bar{\Theta}_i = \bar{\Theta}_{i-1} + \|\vartheta_i + \vartheta_{i-1}\|$
end for

References

1. Vasios, N.; Narang, Y.; Akta, B.; Howe, R.; Bertoldi, K. Numerical analysis of periodic laminar and fibrous media undergoing a jamming transition. *Eur. J. Mech. A/Solids* **2019**, *75*, 322–329.
2. Andrianov, I.V.; Danishevs'kyy, V.V.; Weichert, D. Asymptotic determination of effective elastic properties of composite materials with fibrous square-shaped inclusions. *Eur. J. Mech. A/Solids* **2002**, *21*, 1019–1036. [CrossRef]

3. Godinho, P.M.; Jajcinovic, M.; Wagner, L.; Vass, V.; Fischer, W.J.; Bader, T.K.; Hirn, U.; Bauer, W.; Eberhardsteiner, J.; Hellmich, C. A continuum micromechanics approach to the elasticity and strength of planar fiber networks: Theory and application to paper sheets. *Eur. J. Mech. A/Solids* **2019**, *75*, 516–531. [CrossRef]
4. Tatone, A.; Recrosi, F.; Repetto, R.; Guidoboni, G. From species diffusion to poroelasticity and the modeling of lamina cribrosa. *J. Mech. Phys. Solids* **2019**, *124*, 849–870.
5. Driessen, N.J.; Wilson, W.; Bouten, C.V.; Baaijens, F.P. A computational model for collagen fibre remodelling in the arterial wall. *J. Theor. Biol.* **2004**, *226*, 53–64.
6. Hariton, I.; de Botton, G.; Gasser, T.C.; Holzapfel, G.A. Stress-driven collagen fiber remodeling in arterial walls. *Biomech. Model. Mechanobiol.* **2007**, *6*, 163–175.
7. Driessen, N.J.B.; Cox, M.A.J.; Bouten, C.V.C.; Baaijens, F.P. Remodelling of the angular collagen fiber distribution in cardiovascular tissues. *Biomech. Model. Mechanobiol.* **2008**, *7*, 93. [CrossRef]
8. Grytz, R.; Meschke, G. A computational remodeling approach to predict the physiological architecture of the collagen fibril network in corneo-scleral shells. *Biomech. Model. Mechanobiol.* **2010**, *9*, 225–235. [CrossRef]
9. Humphrey, J.D.; Eberth, J.F.; Dye, W.W.; Gleason, R.L. Fundamental role of axial stress in compensatory adaptations by arteries. *J. Biomech.* **2009**, *42*, 1–8. [CrossRef]
10. Nagel, T.; Kelly, D.J. Remodelling of collagen fibre transition stretch and angular distribution in soft biological tissues and cell-seeded hydrogels. *Biomech. Model. Mechanobiol.* **2012**, *11*, 325–339.
11. Sander, E.A.; Barocas, V.H.; Tranquillo, R.T. Initial Fiber Alignment Pattern Alters Extracellular Matrix Synthesis in Fibroblast-Populated Fibrin Gel Cruciforms and Correlates with Predicted Tension. *Ann. Biomed. Eng.* **2011**, *39*, 714–729. [CrossRef] [PubMed]
12. Crevacore, E.; Di Stefano, S.; Grillo, A. Coupling among deformation, fluid flow, structural reorganisation and fibre reorientation in fibre-reinforced, transversely isotropic biological tissues. *Int. J. Non-Linear Mech.* **2019**, *111*, 1–13. [CrossRef]
13. Baaijens, F.; Bouten, C.; Driessen, N. Modeling collagen remodeling. *J. Biomech.* **2010**, *43*, 166–175. [CrossRef] [PubMed]
14. Irons L.; Humphrey, J.D. Cell signaling model for arterial mechanobiology. *PLoS Comput. Biol.* **2020**, *16*, e1008161. [CrossRef]
15. Engler, A.; Sen, S.; Sweeney, H.; Discher, D. Matrix elasticity directs stem cell lineage specification. *Cell* **2006**, *126*, 677–689. [CrossRef]
16. Discher, D.; Janmey, P.; Wang, Y. Tissue cells feel and respond to the stiffness of their substrate. *Science* **2005**, *310*, 1139–1143. [CrossRef]
17. Chaudhuri, O.; Gu, L.; Darnell, M.; Klumpers, D.; Bencherif, S.A.; Weaver, J.C.; Huebsch, N.; Mooney, D.J. Substrate stress relaxation regulates cell spreading. *Nat. Commun.* **2015**, *6*, 6364. [CrossRef]
18. Chaudhuri, O.; Gu, L.; Klumpers, D.; Darnell, M.; Bencherif, S.A.; Weaver, J.C.; Huebsch, N.; Lee, Ho.; Lippens, E.; Duda, G.N.; et al. Hydrogels with tunable stress relaxation regulate stem cell fate and activity. *Nat. Mater.* **2016**, *15*, 326–334. [CrossRef]
19. Ban, E.; Franklin, J.M.; Nam, S.; Smith, L.; Wang, H.; Wells, R.; Chaudhuri, O.; Liphardt, J.; Shenoy, V. Mechanisms of plastic deformation in collagen networks induced by cellular forces. *Biophys. J.* **2018**, *114*, 450–461. [CrossRef]
20. Wisdom, K.; Adebowale, K.; Chang, J.; Lee, J.Y.; Nam, S.; Desai, R.; Rossen, N.S.; Rafat, M.; West, R.B.; Hodgson, L.; et al. Matrix mechanical plasticity regulates cancer cell migration through confining microenvironments. *Nat. Commun.* **2018**, *9*, 4144. [CrossRef]
21. Buchmann, B.; Engelbrecht, L.; Fernandez, P.; Hutterer, F.P.; Raich, M.K.; Scheel, C.H.; Bausch, A.R. Mechanical plasticity of collagen directs branch elongation in human mammary gland organoids. *Nat. Commun.* **2021**, *12*, 2759.
22. Favata, A.; Rodella, A.; Vidoli, S. An internal variable model for plastic remodeling in fibrous materials. *Eur. J. Mech. A/Solids* **2022**, *96*, 104428.
23. Coleman, B.; Gurtin, M. Thermodynamics with internal state variables. *J. Chem. Phys.* **1967**, *47*, 597–613.
24. Germain, P. The method of virtual power in continuum mechanics. Part 2: Microstructure. *SIAM J. Appl. Math.* **1973**, *25*, 556–575.
25. Halphen, B.; Nguyen, Q. Sur les matériaux standards généralisés. *J. Mécanique* **1975**, *14*, 39–63.
26. Germain, P. Sur certaines définitions liées à l'énergie en mécanique des solides. *Int. J. Eng. Sci.* **1982**, *20*, 245–259. [CrossRef]
27. Germain, P.; Nguyen, Q.; Suquet, P. Continuum thermodynamics. *J. Appl. Mech.* **1983**, *50*, 1010–1020.
28. Favata, A.; Rodella, A.; Vidoli, S. Emerging anisotropy and tethering with memory effects in fibrous materials. *Mech. Mater.* **2024**, *190*, 104928. [CrossRef]
29. Favata, A.; Rodella, A.; Vidoli, S. A variational model for finger-driven cell diffusion in the extracellular matrix. *Meccanica* **2024**, *59*, 1315–1326.
30. Marigo, J.J. Constitutive relations in plasticity, damage and fracture mechanics based on a work property. *Nucl. Eng. Des.* **1989**, *114*, 249–272. [CrossRef]
31. Kuhn, H.W.; Tucker, A.W. *Nonlinear Programming-Proceedings of 2nd Berkeley Symposium*; University of California Press: Berkeley, CA, USA, 1951; pp. 481–492.

32. Kuthe, C.D.; Uddanwadiker, R.V. Investigation of effect of fiber orientation on mechanical behavior of skeletal muscle. *J. Appl. Biomater. Funct. Mater.* **2016**, *14*, 154–162. [CrossRef]
33. Jana, S.; Levengood, S.K.L.; Zhang, M. Anisotropic materials for skeletal-muscle-tissue engineering. *Adv. Mater.* **2016**, *28*, 10588–10612.
34. Yan, J.; Qiang, L.; Gao, Y.; Cui, X.; Zhou, H.; Zhong, S.; Wang, Q.; Wang, H. Effect of fiber alignment in electrospun scaffolds on keratocytes and corneal epithelial cells behavior. *J. Biomed. Mater. Res. Part A* **2012**, *100A*, 527–535.
35. Liu, C.; Zhu, C.; Li, J.; Chai, Y.; Wang, Y.; Fan, W.; Wang, H.; Yan, W. The effect of the fibre orientation of electrospun scaffolds on the matrix production of rabbit annulus fibrosus-derived stem cells. *Bone Res.* **2015**, *3*, 15012. [CrossRef]
36. Petryk, H. Incremental energy minimization in dissipative solids. *Comptes. Rendus Mécanique* **2003**, *331*, 469–474. [CrossRef]
37. Cockburn, B.; Fu, G. A systematic construction of finite element commuting exact sequences. *SIAM J. Numer. Anal.* **2017**, *55*, 1650–1688.
38. Logg, A.; Mardal, K.-A.; Wells, G. (Eds.) Automated Solution of Differential Equations by the Finite Element Method. In *Lecture Notes in Computational Science and Engineering*; Springer: Berlin/Heidelberg, Germany, 2012; Volume 84.
39. Aln, M.S.; Kehlet, B.; Logg, A.; Richardson, C.; Ring, J.; Rognes, E.; Wells, G.N. The FEniCS Project Version 1.5. *Arch. Numer. Softw.* **2015**, *3*, 9–23.
40. Fratzl, P.; Weinkamer, R. Nature's hierarchical materials. *Prog. Mater. Sci.* **2007**, *52*, 1263–1334. [CrossRef]
41. Humphrey, J.D. Review Paper: Continuum biomechanics of soft biological tissues. *Proc. R. Soc. Lond. A* **2003**, *459*, 3–46.
42. Greaves, G.; Greer, A.; Lakes, R.; Rouxel, T. Poisson's ratio and modern materials. *Nat. Mater.* **2011**, *10*, 823–837. [CrossRef]
43. Lakes, R. Foam Structures with a Negative Poisson's Ratio. *Science* **1987**, *235*, 1038–1040. [CrossRef] [PubMed]

materials

MDPI

Article

Numerical Simulation of Compressive Testing of Sandwich Structures with Novel Triply Periodic Minimal Surface Cores

Alexandru Vasile [1,2], Dan Mihai Constantinescu [1,3,*], Andrei Ioan Indreș [1,2], Iulian Constantin Coropețchi [1,2], Ștefan Sorohan [1] and Dragoş Alexandru Apostol [1]

1 Department of Strength of Materials, National University for Science and Technology POLITEHNICA Bucharest, Splaiul Independeței 313, 060042 Bucharest, Romania; alexandru.vasile@mta.ro (A.V.); andrei.indres@mta.ro (A.I.I.); iulian.coropetchi@mta.ro (I.C.C.); stefan.sorohan@upb.ro (Ș.S.); dragos.apostol@upb.ro (D.A.A.)

2 Faculty of Aircraft and Military Vehicles, Military Technical Academy "Ferdinand I", G. Coşbuc Blvd. 39-49, 050141 Bucharest, Romania

3 Institute of Solid Mechanics of the Romanian Academy, Str. Constantin Mille No. 15, 010141 Bucharest, Romania

* Correspondence: dan.constantinescu@upb.ro

Abstract: Sandwich structures with triply periodic minimal surface (TPMS) cores have garnered research attention due to their potential to address challenges in lightweight solutions, high-strength designs, and energy absorption capabilities. This study focuses on performing finite element analyses (FEAs) on eight novel TPMS cores and one stochastic topology. It presents a method of analysis obtained through implicit modeling in *Ansys* simulations and examines whether the results obtained differ from a conventional method that uses a non-uniform rational B-spline (NURBS) approach. The study further presents a sensitivity analysis and a qualitative analysis of the meshes and four material models are evaluated to find the best candidate for polymeric parts created by additive manufacturing (AM) using a stereolithography (SLA) method. The FEA results from static and explicit simulations are compared with experimental data and while discrepancies are identified in some of the specimens, the failure mechanism of the proposed topologies can generally be estimated without the need for an empirical investigation. Results suggest that implicit modeling, while more computationally expensive, is as accurate as traditional methods. Additionally, insights into numerical simulations and optimal input parameters are provided to effectively validate structural designs for sandwich-type engineering applications.

Keywords: TPMS; sandwich structure; finite element simulation; implicit modeling

1. Introduction

Sandwich structures that incorporate TPMS cores are a promising type of lightweight and robust material in the field of mechanical engineering. Characterized by their periodic geometry and minimal surface area, they offer a combination of high strength-to-weight ratio, excellent energy absorption, and adaptability to various design requirements. These structures have attracted significant interest for applications in aerospace, biomedical engineering, and structural mechanics due to their ability to efficiently manage mechanical loads while minimizing material usage. A comprehensive classification of the various types of sandwich structures featuring metamaterial cores is provided in references [1–3].

TPMS topologies, such as the Schwarz primitive, Schoen gyroid, and diamond surfaces, are defined mathematically through equations, enabling precise control over their geometrical parameters and relative density [4,5]. Their mathematical foundation makes them particularly suitable for optimization and integration into sandwich panels, where they act as cores between rigid face sheets. Recent advancements in additive manufacturing have facilitated the fabrication of these complex geometries, enabling fabrication of designs that were previously impractical [6–8].

Experimental studies alone, while providing valuable empirical insights, may not fully capture the complexities of the physical phenomena. Thus, a finite element analysis (FEA) serves as a complementary and indispensable tool for expanding the understanding derived from experiments. In addition, a correspondence between experimental data and FEA increases the confidence in the validity of the obtained results. This is especially important when complex topologies fabricated through additive manufacturing are considered, given the defects such methods could induce, altering their mechanical response [9].

Predicting deformation patterns, stress distribution, and failure mechanisms of TPMS structures under various loading conditions have received extensive interest. Recent relevant studies focus on FEA of the most common types of TPMS structures under compressive [10,11] and impact loading [12,13]. The effects of core geometry, material properties, and relative density on the overall performance of the structures was investigated for parts fabricated through AM of metallic [14,15] or polymeric materials [16,17]. Alternative optimization approaches, such as graded structures [18,19], or hybrid topologies [20,21], have also been proposed. The most commonly used types of FEA software used for simulating TPMS geometries are the *Abaqus* Explicit Solver [22–25] and *Ansys* Explicit Dynamics [26–29].

Implicit modeling allows for the quick generation of different topologies, facilitating the validation of FEA models against experimental data. Consequently, all other design variables are utilized to predict specific mechanical responses, as illustrated in the example described in [30]. To reduce the computational effort of simulating small and complex-shaped TPMS samples, studies tend to analyze only the behavior of a representative volume element (RVE) [26,31]. However, due to the lack of double symmetry and because the general failure mode is influenced by adjacent cells, an analysis of a larger domain is required and often performed [32,33].

This research article focuses on the numerical simulation and analysis of sandwich structures with novel TPMS cores, underlining the benefits of using implicit modeling. A total of eight novel TPMS structures and one stochastic geometry are verified while comparing results with experimental data. The design and fabrication of the topologies were described in more detail in [34]. By presenting computational tools and determining how different input parameters and material models influence FEA results, this study seeks to contribute to the development of lightweight, efficient, and customizable structural solutions for advanced engineering applications.

2. Materials and Methods

Due to the complex shape of the proposed shell-based TPMS samples, the walls of the cells come into contact with each other at high strain values. Thus, an analysis over the entire 10 mm deformation range (33% strain) which was considered experimentally [34] was impossible to perform in the *Static Structural* module of *Ansys*. So, separate analyses were carried out. The first one captured only the sample response in the elastic region and the onset of yielding and was performed in the *Static Structural* module, neglecting any self-contacts and gradual stiffening. The study was continued in the *Explicit Dynamics* module where the samples were simulated up to the deformation considered in the experimental tests, i.e., 10 mm.

2.1. CAD Model for Simulation

Preparing the geometry for FEA is a critical endeavor for an accurate analysis. The initial step involves obtaining a comprehensive representation of the structure, through methods such as computer-aided design, implicit modeling, or three-dimensional scanning. Simulation accuracy depends on the fidelity of the geometry, which ensures reliable results for the investigated physical phenomenon [35].

Computer-aided design (CAD) and implicit modeling are two distinct approaches in geometric representation. CAD employs explicit parametric models with precise geometries defined by vertices, edges, and faces. In contrast, implicit modeling represents surfaces as scalar fields using mathematical functions. This approach is not as widely used in engineering but is more flexible for handling complex shapes and topological changes. Although it has advantages in terms of easy parametrization and visualization of topologies, its usage in FEA models can prove problematic [36]. Different methods have been proposed to reduce the number of triangles used in converting implicit models into CAD-neutral formats [31,37]. In this study, the usage of geometries obtained through both conventional CAD and implicit modeling is presented and compared.

Traditionally, in FEA, CAD-type geometries that are easy to integrate into numerical calculation models are used. This is possible, however, only when the samples are simple enough to be modeled either through parametrization based on mathematical functions or by a geometric approach. The stages of designing a CAD gyroid structure can be seen in Figure 1 and consist of the following: defining the reference tridimensional contour (a) and surface (b), making the necessary translations and multiplying to create the RVE (c), multiplying the RVE according to the proposed dimensions (d), applying the correct thickness of the walls (e), cutting the abnormal faces generated on the sides (f), and adding the face sheets of the sandwich structure (g).

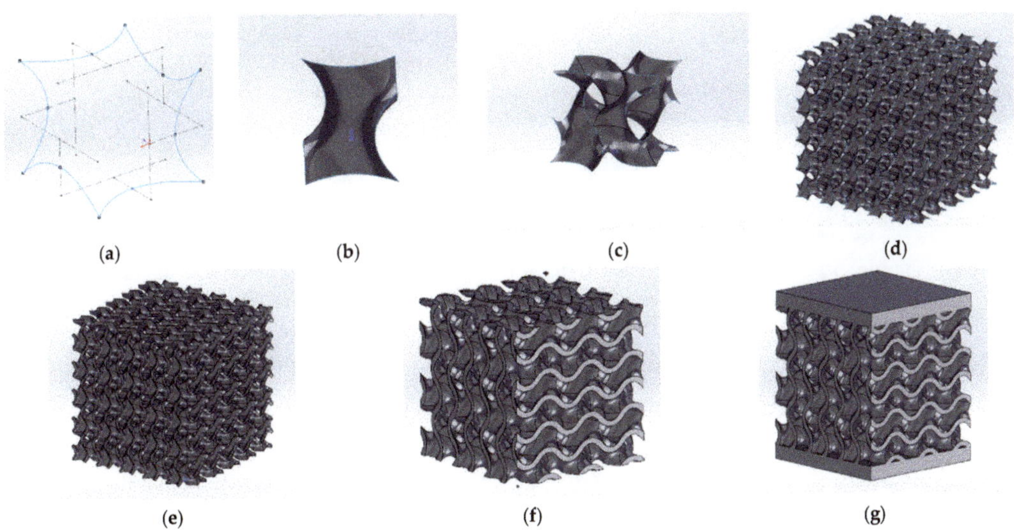

Figure 1. Construction of the CAD-type geometry of sample S1 (gyroid): (**a**) Surface contour; (**b**) Reference surface for 1/8 of RVE; (**c**) The surface of an RVE; (**d**) Multiplication for sample dimensions; (**e**) Obtaining a solid by thickening the walls; (**f**) Cutting to desired dimensions; (**g**) Obtaining the sandwich-type geometry.

The multitude and complexity of these operations help to underline the benefits of using implicit modeling, as all these steps can be replaced by changing the mathematical parameters of the model.

In the case of complex geometries such as topologies S2–S9 defined in [34], modeling by using a CAD approach becomes problematic. When the functions used are complicated, the transition from the explicit form of the defining function $f(x, y, z) = c$ to the parametric equations $x = f(t)$, $y = g(t)$, $z = h(t)$, used by conventional design software, is difficult. In Figure 2, the procedure that resulted from implicit modeling used in this paper to generate a complex model integrated into numerical analyses is proposed.

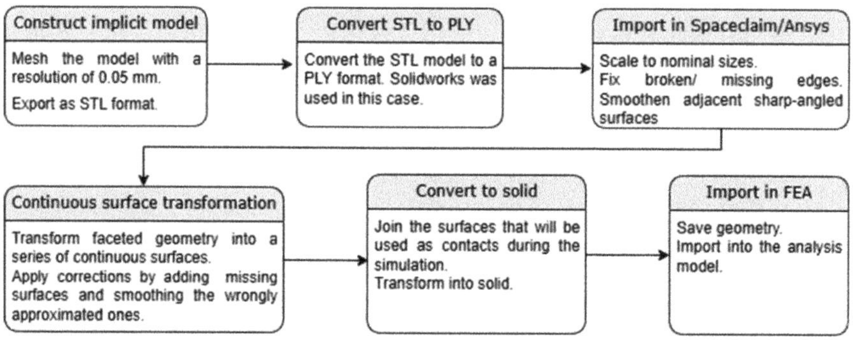

Figure 2. Procedure for obtaining solids usable in finite model analyses starting from implicit modeling.

This was achieved through the following steps. First, the implicit topology was created in the *nTopology* (version 4.5.3) software. The implicit volume was meshed with a size of 0.05 mm, and the resulting geometry was exported in the STL format. This dimension was adopted because it is the same as the layer height that was used in the AM process. Then, the STL format is converted into a PLY format. We used *Solidworks 2023* software, as our observations indicated that this model produces a faceted body with fewer errors, requiring simpler further processing. The resulting format is then imported into the *Spaceclaim* module of *Ansys Workbench 2024 R1*. Because the PLY format does not retain information related to the dimensions of the sample, it is necessary to scale the imported part. This was initially performed for the gyroid geometry, for which the CAD format is also available. Samples being of the same size, the conversion factor remains the same for each subsequent sample. A faceted-surface-type geometry is thus obtained, consisting of surfaces specific to the mesh previously created. The next step is the correction of the discontinuous or missing edges and the smoothing of adjacent surfaces with sharp angles between them. This is necessary because the continuity errors appearing on the surface of the samples make the successful creation of a solid impossible. The geometry is transformed into higher-dimensional surfaces using the *Autoskin* feature. This transformation can generate approximation errors, leading to missing surfaces or erroneous edges. It was therefore necessary to fix the geometry by simplifying the sides of the surfaces, smoothing the wrongly discretized sides, removing the extra surfaces, joining the small surfaces, and filling the gaps. One step that simplifies the subsequent application of the boundary conditions is to join the surfaces to be used as contacts in the FEA, such as the top and bottom sheets of the sandwich structure. After this step, the conversion to a solid is performed, which can be further imported as geometry into the numerical analysis model.

2.2. Material Definition

The second challenge consists in defining, as accurately as possible, the behavior of the material used to manufacture the samples. Although the library of the *Mechanical* module of *Ansys* contains a large number of predefined material models, it does not address the majority of materials used in AM processes, especially SLA. Thus, to ensure the accuracy of the model, the mechanical properties of the commercially available photosensitive resin used (Tough 1500 v1—Formlabs—Somerville, MA, USA) were specified based on experimental data obtained from uniaxial tensile tests.

Ten samples were fabricated using the same printing parameters and evaluated under uniaxial loading. Tensile tests were performed, according to the ASTM D638 standard [38], on an Instron 68TM-50 (Instron—Norwood, MA, USA) static and dynamic testing system with a 10 kN force cell and an Instron 2630-100 (Instron—Norwood, MA, USA) digital extensometer. Figure 3 briefly portrays the fabrication and testing setup of the tensile specimens. The conventional stress–strain curves were subsequently converted to true stress–strain curves.

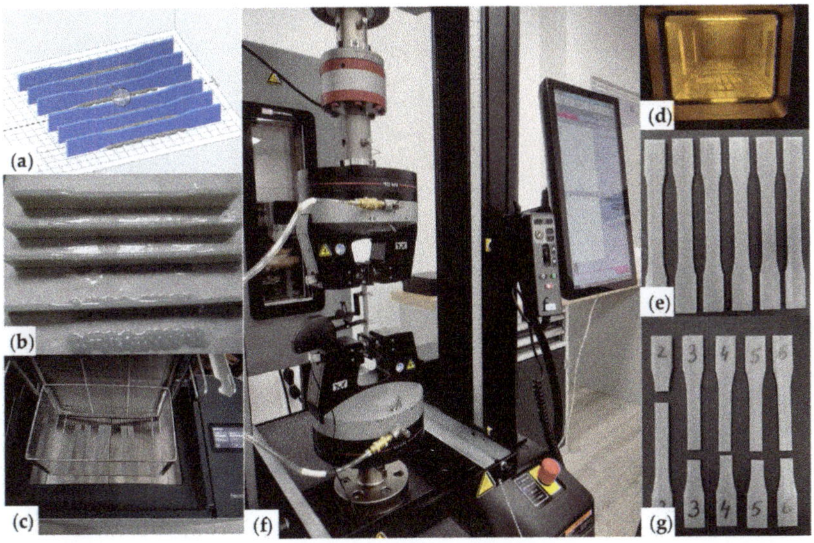

Figure 3. Fabrication of tensile samples and their uniaxial tensile tests: (**a**) Layout of the samples in the printing software; (**b**) Samples on the printing table; (**c**) Post-processing by washing with isopropyl alcohol; (**d**) Post-processing by treatment with ultraviolet radiation; (**e**) Printed samples; (**f**) The uniaxial testing configuration of the specimens; (**g**) Specimens after tensile tests.

The red curve represented in Figure 4 shows the average (mean) true stress–strain curve for the 10 uniaxial tensile tests carried out. An initial region of linear elastic behavior of the material is observed up to a strain of 2%. After yielding, a strain-softening tendency is identified, where the slope of the curve becomes negative, and a necking zone can be observed on the specimen. The third region corresponds to a long hardening region, specific to polymeric materials, with an almost linear evolution until the final rupture of the specimen.

To use these curves as a reference behavior for numerical simulations, a separation between the linear elastic and non-linear behavior of the material is needed. Considering, based on the analysis performed on the homogeneity of the samples [39], that the material has isotropic properties, the first region was defined considering the mean value for the longitudinal (Young's) modulus of elasticity of E = 1296 MPa and a Poisson's ratio ν = 0.35, based also on the observations from relevant literature [40,41]. Next, using four material models presented below, the non-linearity of the material was approximated.

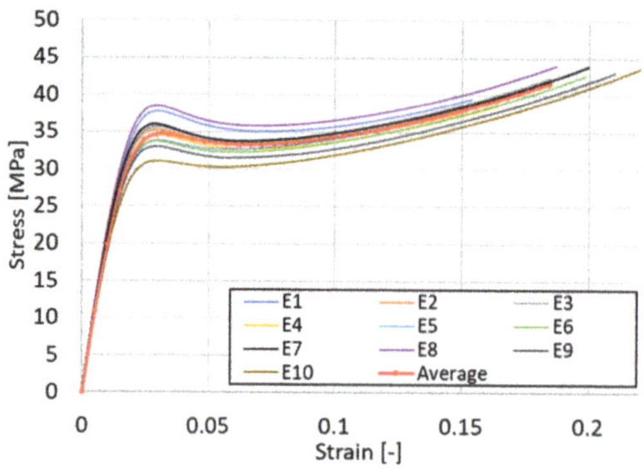

Figure 4. True stress–strain plots obtained through tensile testing of 10 specimens and the mean curve.

The simplest approach is to use a bilinear approximation of the curve, where the initial linear segment given by the slope E characterizes the elastic region, while the subsequent linear segment accounts for the hardening behavior after yielding and is given by the tangent modulus E_t. The relatively constant hardening rate after yielding makes the model suitable for the studied polymeric material. The bilinear isotropic hardening model that approximates the experimentally obtained data is presented in Figure 5a, with the defining parameters given in Table 1.

Table 1. Parameters used to define the material models used in FEA.

Bilinear isotropic hardening model		**Multilinear isotropic hardening**		
Young's modulus E = 1296 MPa		$\varepsilon_{el} = \frac{\sigma_{real}}{E}$		(1)
Yield strength σ_y = 32.12 MPa		$\varepsilon_{pl} = \varepsilon_{real} - \varepsilon_{el}$		(2)
Tangent modulus E_t = 90.12 MPa		$\sigma_{real} = 138{\varepsilon_{pl}}^2 - 6.7\varepsilon_{pl} + 35$		(3)
Power law non-linear isotropic hardening		**Three-network model [24]**		
$\frac{\sigma_{real}}{\sigma_y} = \left(\frac{\sigma_{real}}{\sigma_y} + \frac{3G}{\sigma_y}\varepsilon_{pl}\right)^N$		Shear modulus A—140.5 MPa	Shear modulus evolution rate B—16	
		Flow resistance A—51.5 MPa	Flow resistance B—76.5 MPa	
	(4)	Stress eponential A—1.4	Stress eponential B—2.1	
N = 0.15 [41]		Initial shear modulus B—362.5 MPa	Shear modulus C—2.5 MPa	
		Final shear modulus B—22.5 MPa	Locking stretch—16 Bulk modulus—502 MPa	

A model that overlays the experimental curve with better accuracy is the *multilinear hardening plasticity*, which does not assume a constant slope and allows a more realistic representation of the material response by incorporating several linear segments. Its definition was achieved by introducing several points corresponding to the true stress–true plastic strain curve. Each segment corresponds to a distinct hardening stage, allowing the model to capture the evolving behavior of the material as it undergoes successive plastic deformations. Since the model does not allow for defining a negative slope, which is visible in the red curve presented in Figure 4, it was necessary to convert the true stress–strain evolution into a curve that is increasing continuously. The construction of a polynomial function that superimposes as much as possible over the real variation, described by relation (3) in Table 1, was achieved and is represented in Figure 5b. The other formulas used in defining the model of material are relations (1) and (2), where: σ_{real}—true stress value, ε_{pl}—plastic strain, ε_{el}—elastic strain, E—Young's modulus.

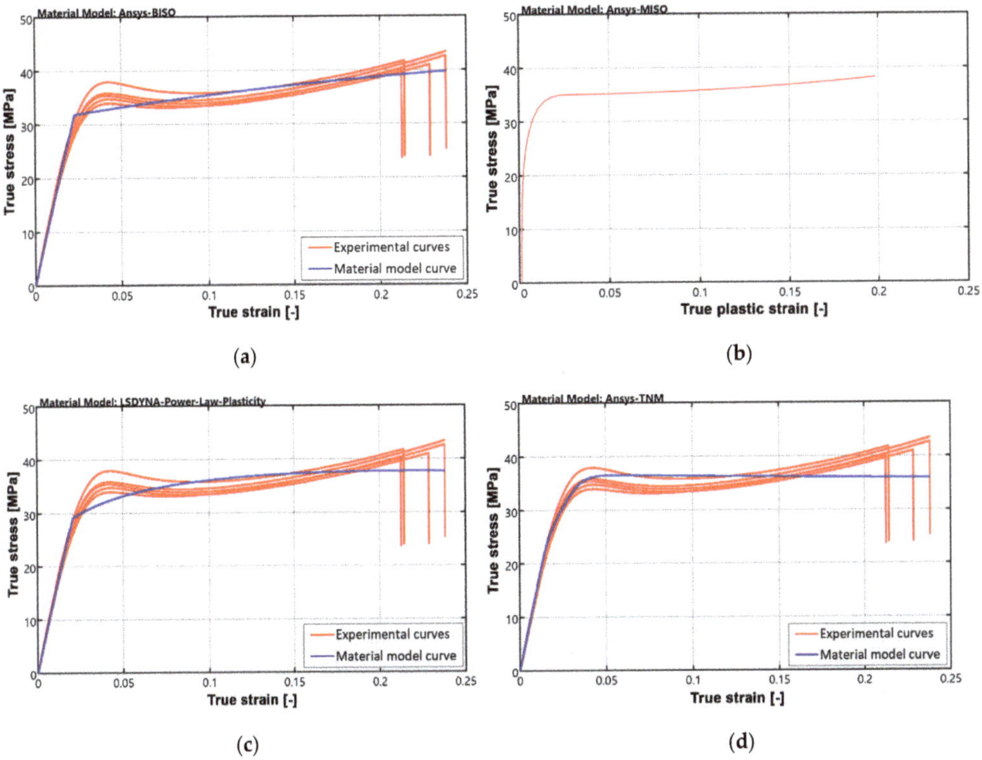

Figure 5. Material models proposed: (**a**) bilinear isotropic hardening; (**b**) multilinear isotropic hardening; (**c**) power law plasticity; (**d**) three-network model.

Considering the numerous material models available in FEA libraries used for numerical simulations and the complexity involved in defining them, it is beneficial to adopt simplifying approaches when working with more complex materials. Such a model is the *power law non-linear isotropic hardening* model, used in the case of materials that exhibit a non-linear hardening in the plasticity zone, defined by (4), where: σ_y—yield stress value, G—transversal modulus of elasticity, and $N = 0.15$ constant exponent as defined in [42]. In Figure 5c, the resulting model curve is plotted together with the experimental data.

Working with any material model requires determining appropriate material parameters from experimental data. A good option is the use of the package *PolyUMod*® (*Polymer FEM*, bought by *Ansys*) whose family of material models has incorporated advanced user-material models for finite element modeling of polymers, biomaterials, and other non-linear materials. To facilitate parameter extraction, *Polymer FEM* has developed *MCalibration*, which enables the semi-automatic extraction of pertinent material parameters from experimental data. This allows for the use of an iterative approach, in which the parameters and coefficients from the predefined material model formulas are modified until its curve approximates as accurately as possible the true strain–stress curves obtained experimentally by testing the tensile specimens. These parameters were then entered into the *Ansys* software *Workbench 2024 R1*, into the appropriate models, to be used in the FEA.

Due to the observations made after compressive testing, when the samples returned almost to their original shape within a couple of minutes, a material model that takes into account a viscoelastic behavior was also considered. The *three-network model (TNM)* states that three time-dependent viscous networks in molecular equilibrium give the mechanical response of the material, acting in parallel. In Table 1, the parameters that define the model, customized within *PolyUMod*, are manually adjusted and presented following the parameter notations found in *Ansys*. Figure 5d highlights the resulting curve. Volghin and Shishkovsky provide similar examples of the use of the model in [43] and Kumar in [44], for simulating the response of a part made from polylactic acid. The advantage of the model is that it allows capturing specific softening curves with negative slope that usually occur in polymers after yielding (Figure 5d).

2.3. Geometry Meshing

Another element that fundamentally influences the accuracy and reliability of FEA is the realization of a mesh adapted to the geometry and loading conditions, which ensures the faithful representation of the structural response [45]. The fineness, element type, and method used for meshing the model directly influence the convergence of the solution and the ability to capture the failure mode of structures with intricate topologies.

In the case of the proposed TPMS samples, although the CAD-type geometry could be meshed with quadratic elements, the models designed through implicit modeling presented errors or significantly reduced values for the mesh quality when trying to use the same order of elements. For this reason, to maintain consistent results, it was decided to discretize all the models using tetrahedron elements. The solid parts were discretized using Solid187 elements, while the contact area was constructed using Targe170 elements, which are specific to rigid materials and therefore applicable to the platens of the testing machine and Conta174 elements, which are specific for deformable elements and therefore applicable for the geometry of the metamaterial [46,47].

Creating a matching mesh between the elements of both surfaces in contact is crucial for accurately capturing their mechanical behavior. If this is easily obtained for the upper and lower surfaces of the specimen given by the contact between the platens and the face sheets of the sandwich structures, achieving compatibility between the elements of the metamaterial walls that come into contact with each other is difficult to realize. This is one of the reasons why convergence is often not reached at greater strain values.

Several meshing methods were verified to see which one works better for the proposed geometries. The simplest and most effective is the "*Automatic Method*". However, it generated meshing errors in certain regions, consisting of several surfaces with large radii of curvature. The complicated nature of the topologies made the "*Multizone*" method exhibit serious implementation difficulties and unrepeatability between samples when separate meshing of the face sheets and the metamaterial core was attempted. The "*Patch Conforming*" method is well-suited for complex geometries, providing enhanced control over mesh quality. It enabled the successful meshing of all analyzed samples.

A convergence study was carried out separately for the geometry of the test model assembly. The metamaterial solid was meshed with elements with a maximum size of 0.3 mm, 0.5 mm, 1 mm, 2 mm, or 3 mm, while the compression machine platens were discretized into two regions using spheres of influence, visible in Figure 6a. The minimum dimension of 0.3 mm was selected to have at least 3 elements along the thickness of the sample walls. The platen region in contact with the sample and its immediate vicinity was meshed with elements whose size coincided with that of the tested part. In the outer area of the platens the size was increased up to 10 mm.

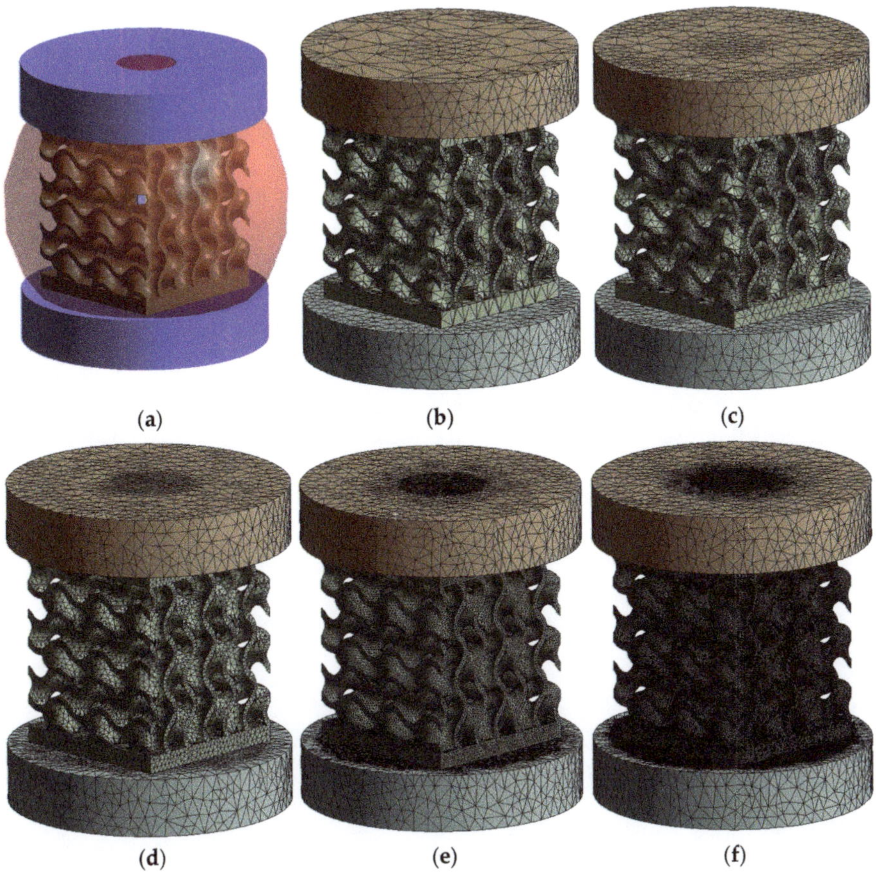

Figure 6. (**a**) Geometry of the gyroid model; (**b**) 3 mm maximum mesh element size; (**c**) 2 mm maximum mesh element size; (**d**) 1 mm maximum mesh element size; (**e**) 0.5 mm maximum mesh element size; (**f**) 0.3 mm maximum mesh element size.

To indicate the mesh quality, the "*Element quality*" index available in *Ansys* was used, where values close to 1 show elements formed from equilateral tetrahedrons, while values close to 0 indicate deformed elements with sharp angles. The latter introduces large errors in capturing the physical phenomenon. Figure 6b–f show the different wall thickness meshes considered for the sensitivity analysis.

2.4. Boundary Conditions

For the first static analysis, the *Static Structural* module of *Ansys* that aimed to capture the mechanical response in the elastic deformation zone until the onset of yielding is used. The following boundary conditions were considered: the contact between the platens of the testing machine and the faces of the sandwich structure was frictionless, the lower platen had all its degrees of freedom blocked, and a 2.5 mm vertical displacement was applied to the upper platen.

The second numerical simulation was carried out in the *Explicit Dynamics* module, where it was possible to take into account the self-contact that occurred between the walls of the samples over the entire 10 mm deformation range. In this case, to simplify the model, a 10 mm deformation was introduced to the upper surface of the TPMS structures, without considering in the model the platens of the testing machine. The bottom surface was considered fixed.

3. Results

Next, the results obtained from numerical simulations are presented. As previously mentioned, the geometries obtained through implicit modeling require an increased computational effort and a more complex meshing procedure. In order to gain a detailed understanding of the phenomenon and establish how the parameters used in the FEA model influence the results, several tests were carried out on the gyroid geometry, considered as a benchmark topology.

3.1. Influence of Mesh Size

Initially, a convergence study was performed in order to analyze how the maximum size of the elements influence the final results. Figure 7a displays the percentage distribution of the number of elements of each mesh. Figure 7b shows the meshing qualitative parameters. It is observed that, contrary to expectations, there is no proportionality between the maximum size of the mesh and the number of elements into which the solid is discretized. This is due to the large elements generated in areas with large radii of curvature, which require a finer mesh in adjacent areas where the radius of curvature of the surfaces decreases. However, a positive trend in the average quality index and a negative trend in the standard deviation are observed as the refinement of the mesh increases. Also, the number of elements along the thickness of the geometry walls is important, noting that in the case of a mesh with an element size greater than 1 mm, a single element was generated throughout the thickness. This resulted in element shapes that significantly deviate from the ideal equilateral triangle shape typical of regular tetrahedron elements. As the mesh size is decreased, the walls are divided into more layers, which facilitates the creation of higher-quality elements along the surfaces.

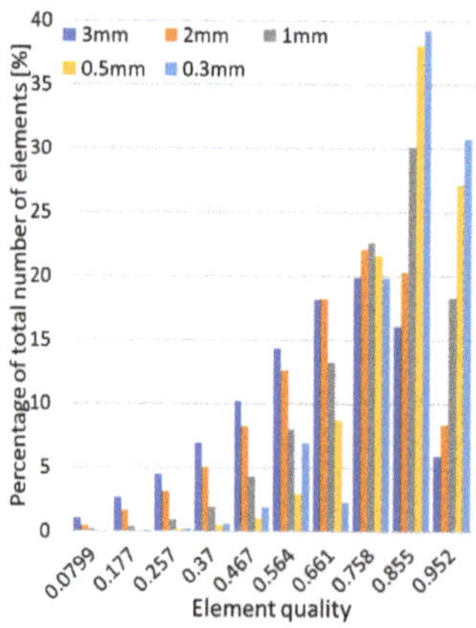

(a)

Meshing qualitative parameters

Maximum Mesh Size [mm]	No. of Elements	Average Quality Index	Standard Deviation
3	263,278	0.645	0.194
2	351,552	0.686	0.181
1	374,898	0.770	0.152
0.5	2,207,825	0.833	0.107
0.3	9,991,715	0.844	0.102

(b)

Figure 7. (**a**) Meshing quality index distribution; (**b**) Meshing qualitative parameters.

An analysis of this type is useful because it became obvious that when the mesh refinement increased, the percentage of elements of higher qualities also increased significantly.

Thus, the total number of distorted elements has been considerably reduced, leading to a lower probability of convergence issues of significant errors for the proposed gyroid model.

Figure 8 shows the compressive force curves as a function of deformation for different mesh sizes. The study was carried out in the *Static Structural* module of *Ansys Mechanical*, on the gyroid geometry, not taking into account the self-contacts that occur during deformation. The material selected was the *Multilinear* model presented in Figure 5.

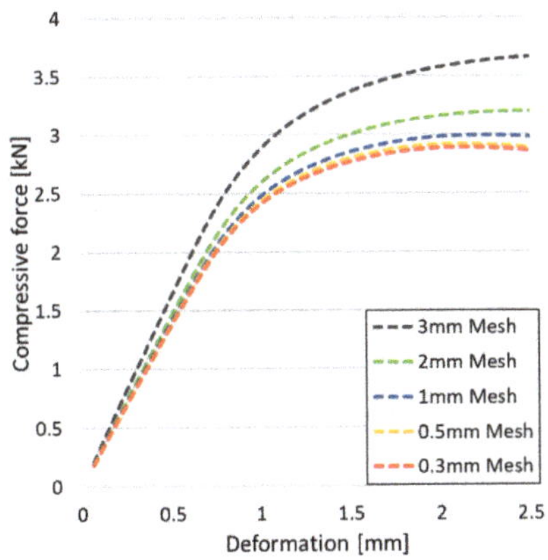

Figure 8. Evolution of the compressive force as a function of deformation for different mesh sizes.

As the mesh refinement is increased, the value of the compressive force decreases significantly. If the model is composed of more elements, the value of the loading force is calculated more precisely. This is based on the increased ability to capture stress gradients more accurately across the part. There is, however, a point of diminishing improvement, where only computational time was added without a significant increase in model accuracy. For this reason, a mesh with a maximum size of 0.3 mm was adopted in order to provide a reasonable balance between convergence time and result accuracy. This mesh size was kept constant in all subsequent analyses.

3.2. *Influence of Material Models*

Due to the typical behavior of the polymer material, given by a higher yield stress followed by a softening region and stiffening at significant deformations, several material models were proposed to establish which characterizes the material behavior with the best accuracy.

Figure 9a shows the compressive force as a function of the deformation for the numerical simulations (dotted lines) obtained for the gyroid sample with the material models described in Figure 5. It is observed that all of them give a response close to the average experimental curve obtained through compressive testing of three identical samples (red curve). All models present a similar stiffness deviation in the elastic region but follow different yield responses. In order to correctly estimate which of the four models came closest to the experimental data, the difference between the area under the curve of the numerical simulations for each model and the area under the curve of the average force recorded empirically was evaluated. Figure 9b highlights this deviation. Based on this plot, deviations of 2.54% for the *Multilinear* model, 4.65% for the *Bilinear* model, 5.08% for the *Powerlaw* model, and 3.16% for the *TNM* were determined.

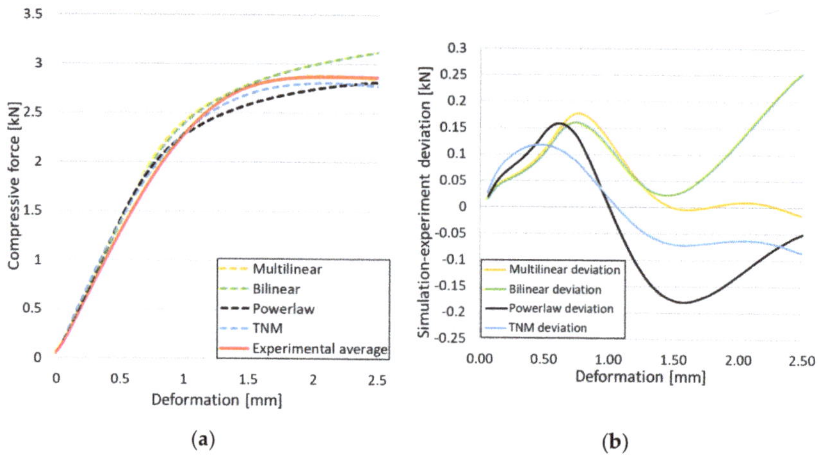

Figure 9. (**a**) Compressive force as a function of deformation for the proposed material models; (**b**) Difference between the dotted curves in (**a**) and the average experimental data (red curve).

Thus, the *Multilinear* model presented the highest degree of curve overlap, most accurately capturing the real behavior of the samples, and was further used in the FEA.

3.3. Influence of Using CAD-Type Geometry or Geometry Obtained by Implicit Modeling

Considering that most of the built samples cannot be reproduced by conventional CAD means, the problem that arises is whether the use of the geometry obtained through implicit modeling introduces any errors in the FEA. For this purpose, a comparison was performed, based on the gyroid sample as a benchmark test, between the results obtained when the solid used in the numerical simulations was a conventional CAD geometry or was obtained as indicated in Figure 2. The test conditions are the same as previously stated, the only variable being the geometry type.

Figure 10 visually shows the two formats used, the CAD one in (a), defined by the procedure presented in Figure 1, and the one obtained after correcting the geometry obtained through implicit modeling in (b), following the procedure described in Figure 2. Figure 10c displays the geometric deviations between the two models, with an imposed tolerance of 0.1 mm. This provides an overview on the fidelity of reproducing the nominal geometry through implicit modeling in order to see how these deviations may generate different outcomes in the numerical simulation.

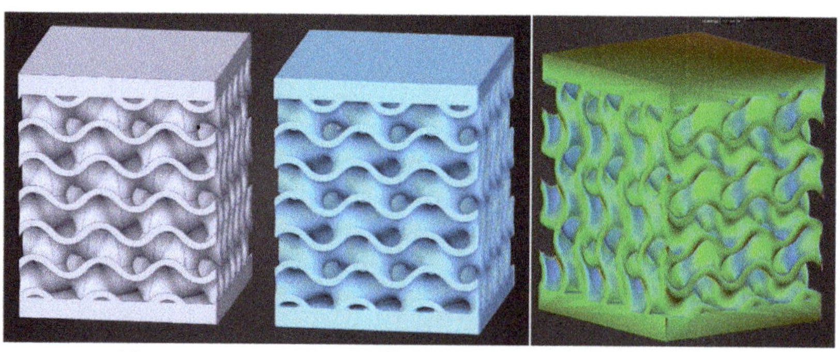

Figure 10. (**a**) CAD geometry; (**b**) Solid obtained from implicit modeling; (**c**) Dimensional deviations between the two formats.

Figure 11 displays the differences obtained by using the two types of geometries in the computational model. For validation, all the material models presented above were verified. A very good overlap between the two pairs of tests is observed. The conclusion is that utilizing a mesh of 0.3 mm to convert the implicit model into a solid compatible with numerical simulation yields results that closely match those obtained through conventional analysis using CAD geometries.

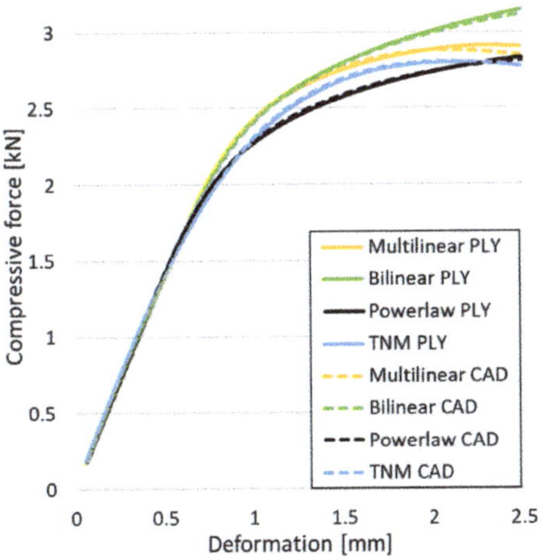

Figure 11. Compressive force as a function of deformation for CAD-type geometries or obtained through implicit modeling.

3.4. *Finite Element Analysis in Ansys Static Structural Module*

The analysis of the results obtained by numerical simulation in the *Ansys Static Structural* module is presented below. Figure 12 presents the meshed geometric model and a comparison between the average force–deformation curve obtained from three experimental tests performed on each proposed structure and the curves obtained from FEA. An approach similar to that presented in Figure 9b was used to identify the degree of overlap of the curves, obtaining the following relative deviations from the empirical data: S1—1.54%, S2—13.78%, S3—10.61%, S4—7.61%, S5—8.56%, S6—29.11%, S7—8.5%, S8—1.8%, S9—30.3%, S10—10.53%. It can be seen that all samples exhibited higher FEA stiffnesses in the elastic region, which can be attributed to differences between the longitudinal modulus of elasticity obtained from the tensile tests and its simulated value. A tendency of higher simulation error from the experimental results is observed as the yield strength of the samples decreases. If geometries S1 and S8 present the highest compressive forces and the lowest deviations, at the opposite side are samples S6 and S9 with the lowest accuracies and the lowest capable forces. The high discrepancy indicated in the case of sample S6 can be attributed to the small features of the topology and the smaller wall thickness. This did not allow for three elements to be generated across the walls of the entire cell, thus reducing the accuracy of the results. In the case of specimen S9, the reason behind the high error could be its deformation mode. Due to its inclined layers, the structure does not show a bending-dominated behavior, such as for the other samples. An explicit analysis of this specimen could provide better results. At the same time, the discrepancies could also be related to alterations generated during the AM printing.

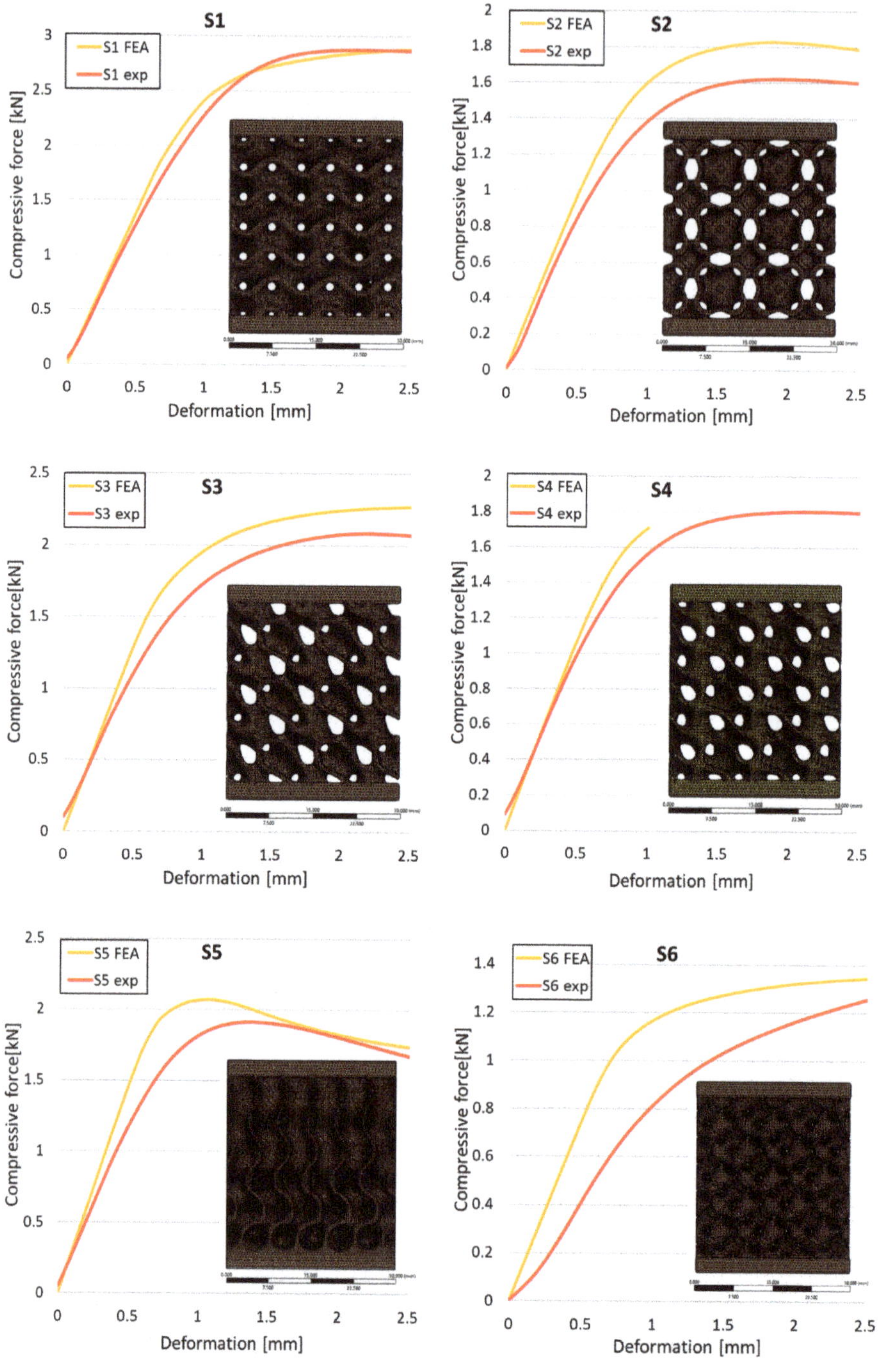

Figure 12. *Cont.*

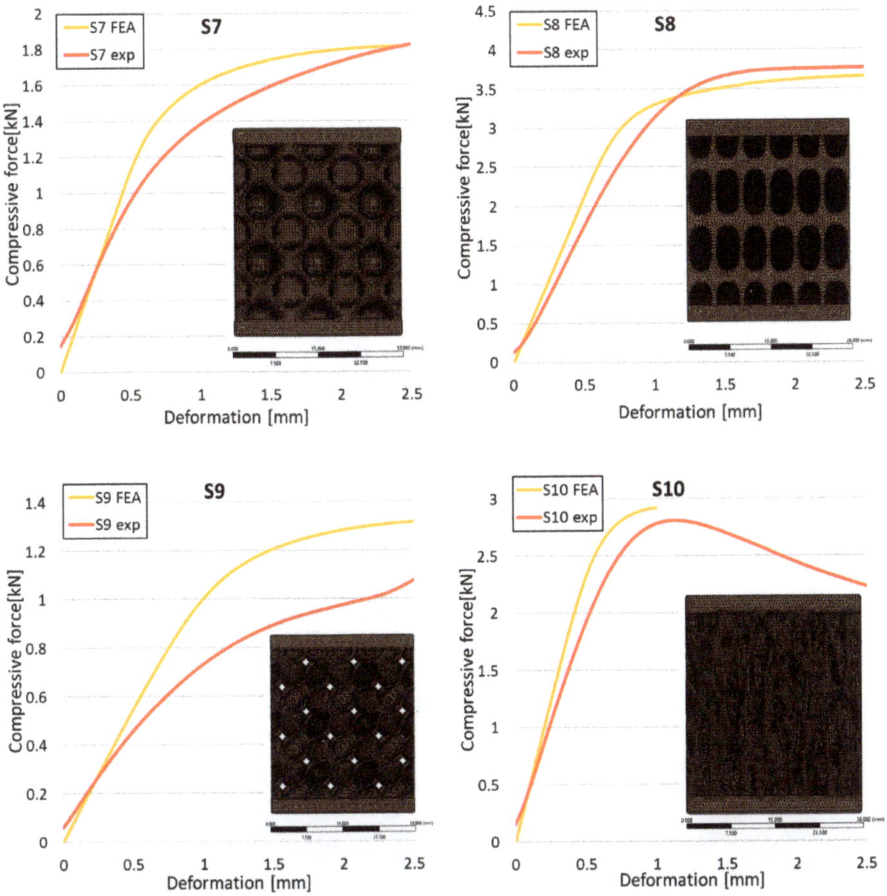

Figure 12. Comparison between the force–deformation curves obtained experimentally and by finite element analysis for each geometry.

The equivalent von Mises stress criterion is specific to ductile materials such as metals. Polymers, especially under compressive loads, can exhibit complex behaviors such as viscoelasticity, non-linearity and time-dependent deformation, introducing uncertainties in the application of the criterion [48]. However, in many cases of testing parts made from plastic polymers by AM, especially for quasi-static loading, von Mises stresses are intensively used and can serve as an approximate criterion for identifying the yield phenomenon [11,27,31]. This is particularly important, especially for complex structures, where the stress states can vary significantly due to geometric complexity and three-dimensional stress distributions. Table 2 presents the distribution of equivalent von Mises stresses and equivalent total strain for the static simulations performed, along with a representation of the experimental compressive response of each structure. The yellow contour in the last column highlights the areas where visible deformation was first observed during experimental testing. The red curve indicates the failure contour of the specimen.

Table 2. Visualization of FEA results and experimental compressive response for each sample.

Sample (C1)	Von Mises Equivalent Stress Distribution [MPa] (C2)	Total Strain Distribution [-] (C3)	Compressive Response (C4) [34]
S1 (R1)	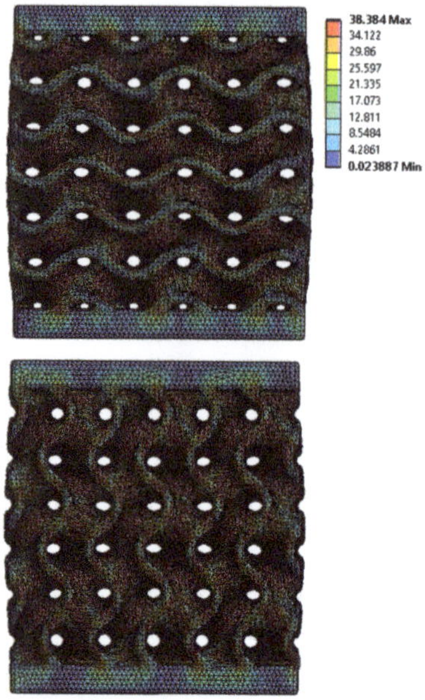	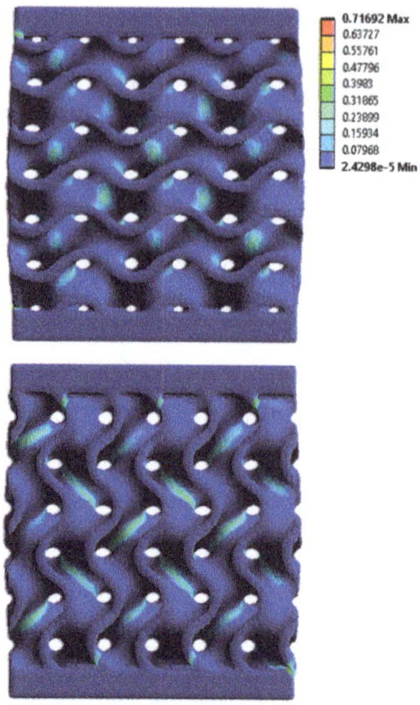	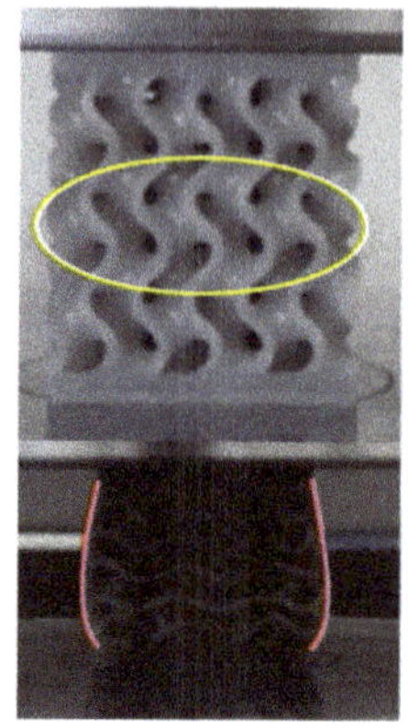
S2 (R2)	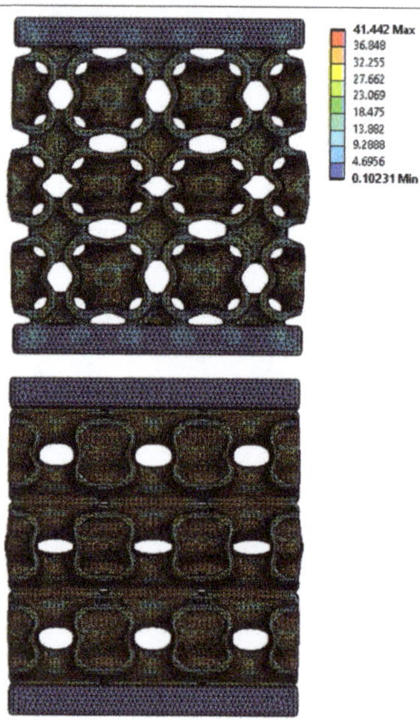	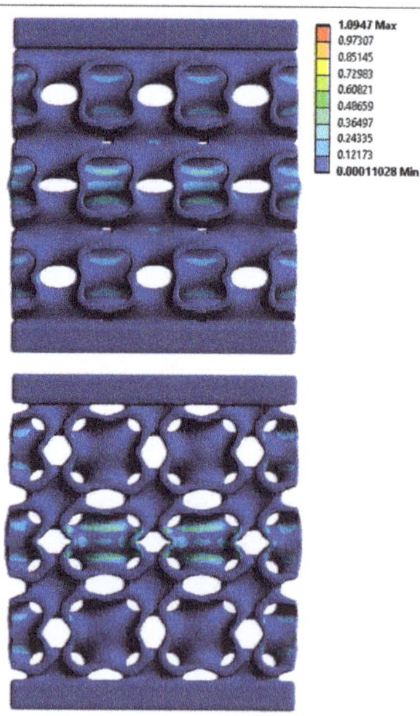	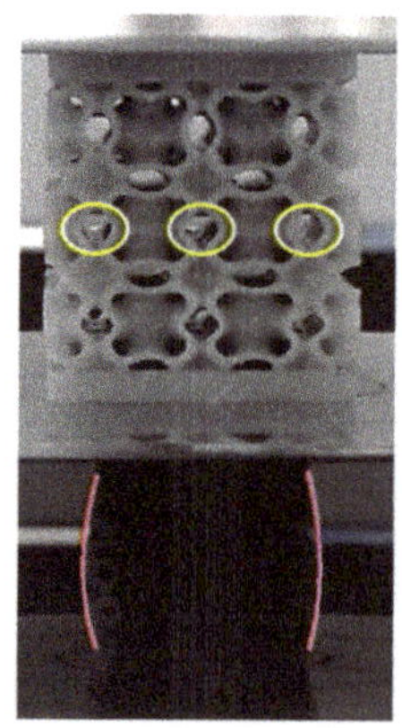

Table 2. *Cont.*

Sample (C1)	Von Mises Equivalent Stress Distribution [MPa] (C2)	Total Strain Distribution [-] (C3)	Compressive Response (C4) [34]
S3 (R3)	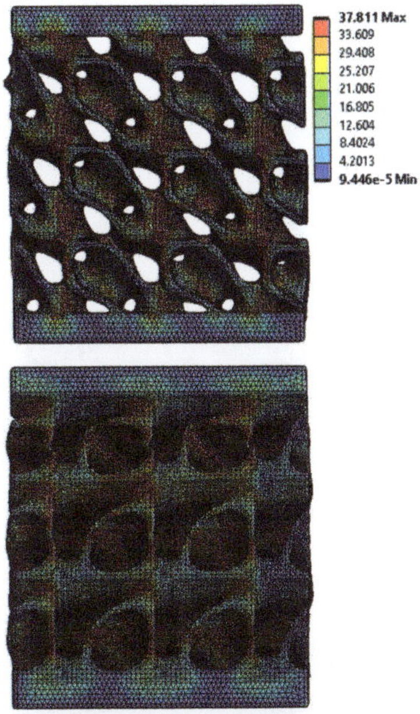	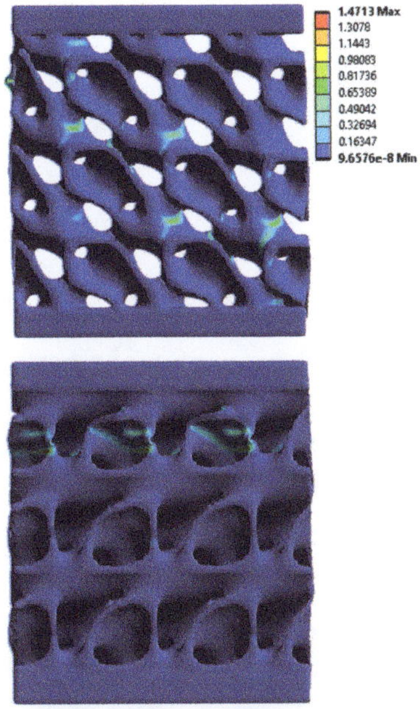	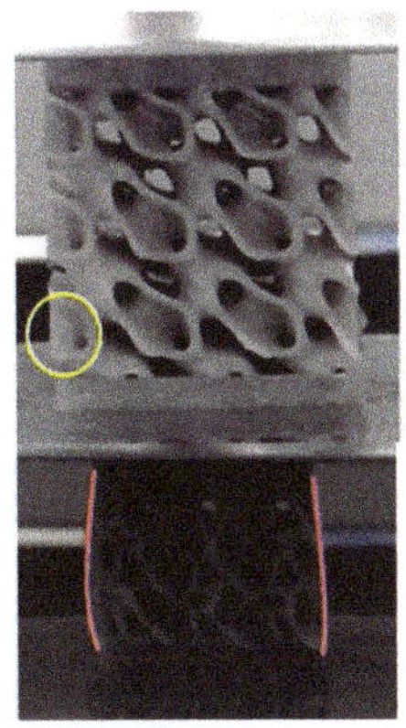
S4 (R4)	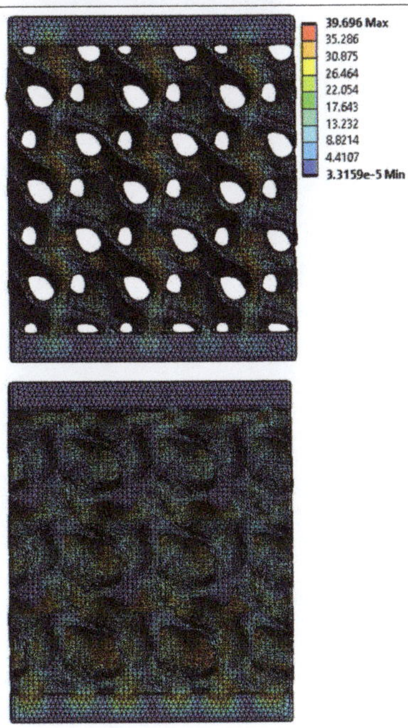	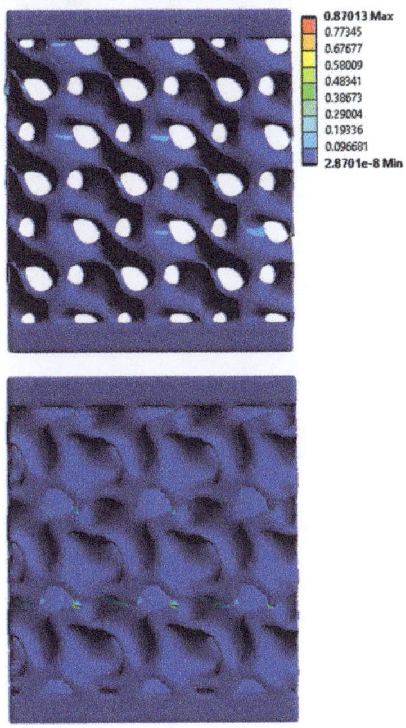	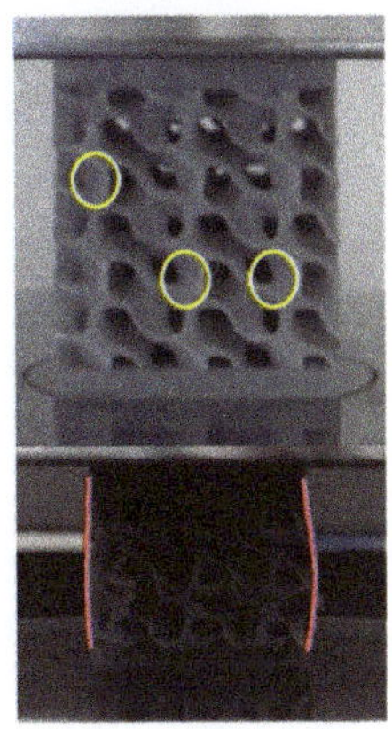

Table 2. *Cont.*

Sample (C1)	Von Mises Equivalent Stress Distribution [MPa] (C2)	Total Strain Distribution [-] (C3)	Compressive Response (C4) [34]
S5 (R5)		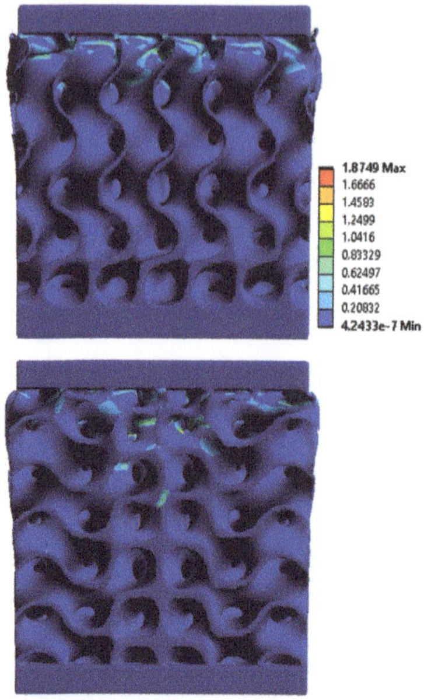	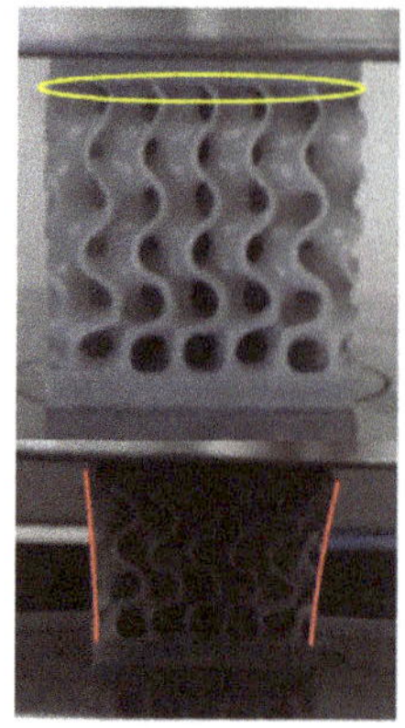
S6 (R6)	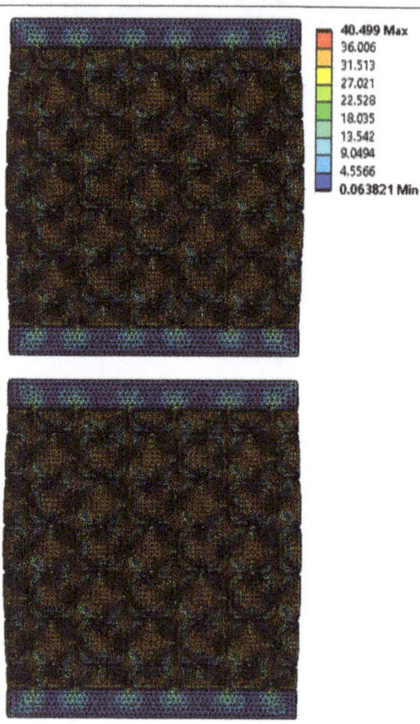	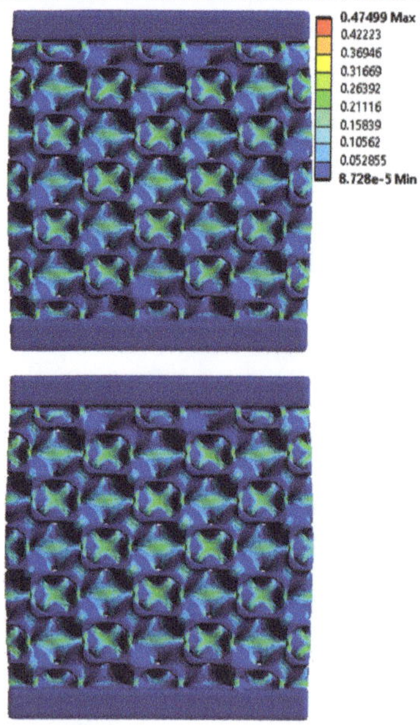	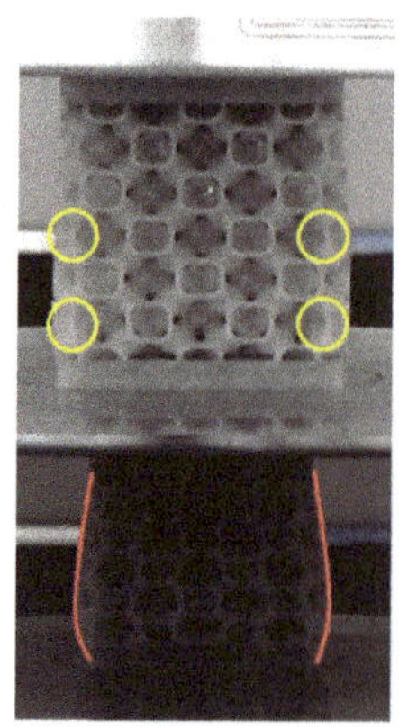

Table 2. *Cont.*

Sample (C1)	Von Mises Equivalent Stress Distribution [MPa] (C2)	Total Strain Distribution [-] (C3)	Compressive Response (C4) [34]
S7 (R7)	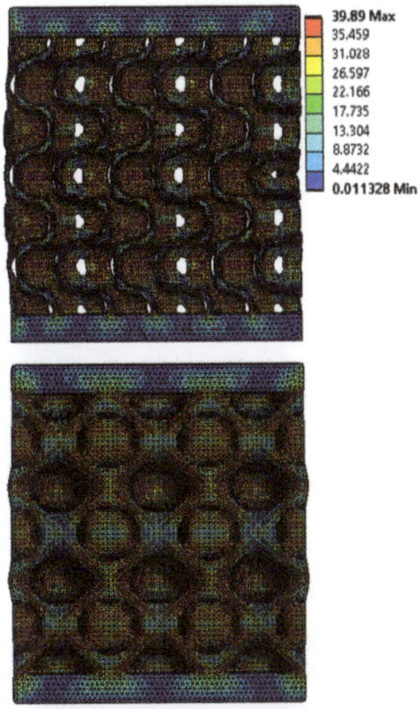	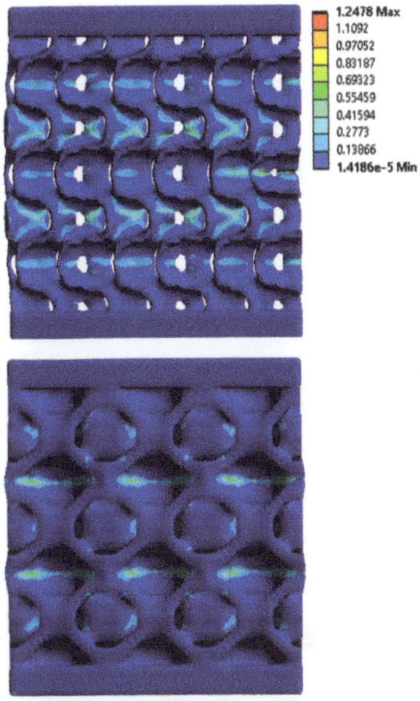	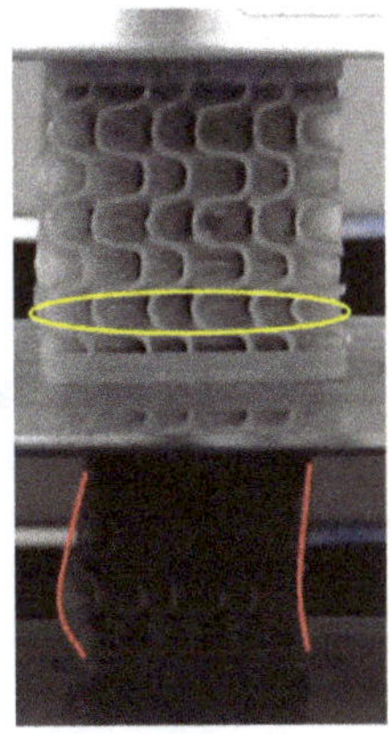
S8 (R8)	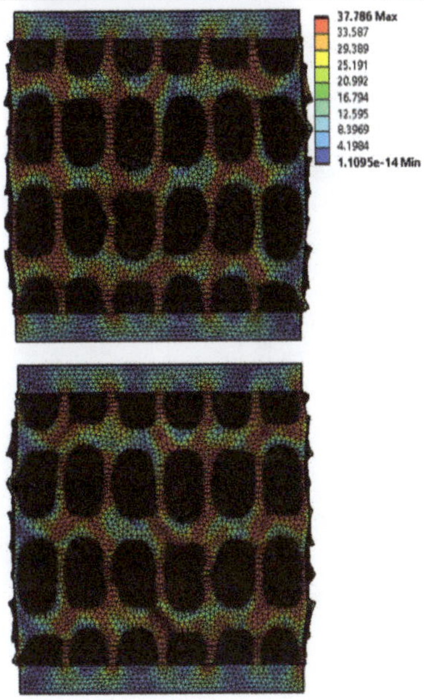	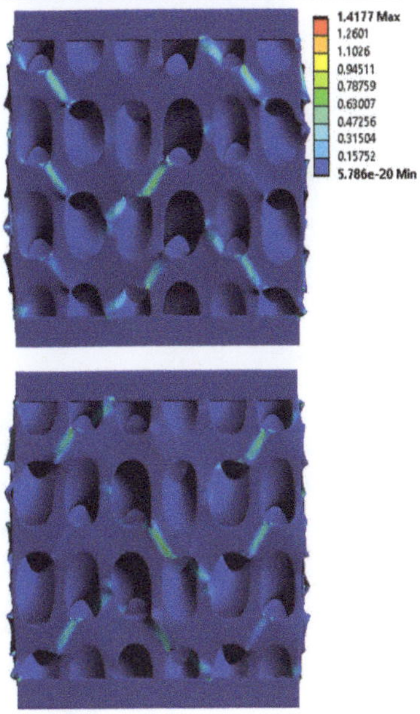	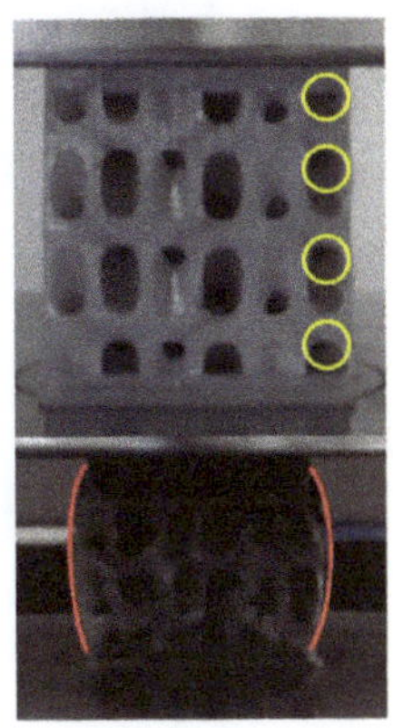

Table 2. *Cont.*

Sample (C1)	Von Mises Equivalent Stress Distribution [MPa] (C2)	Total Strain Distribution [-] (C3)	Compressive Response (C4) [34]
S9 (R9)	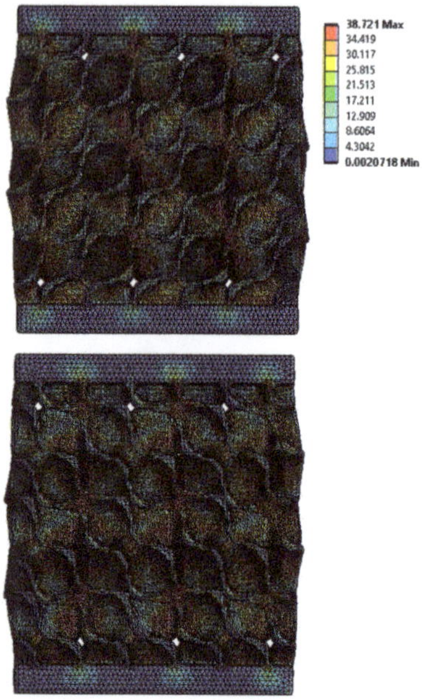	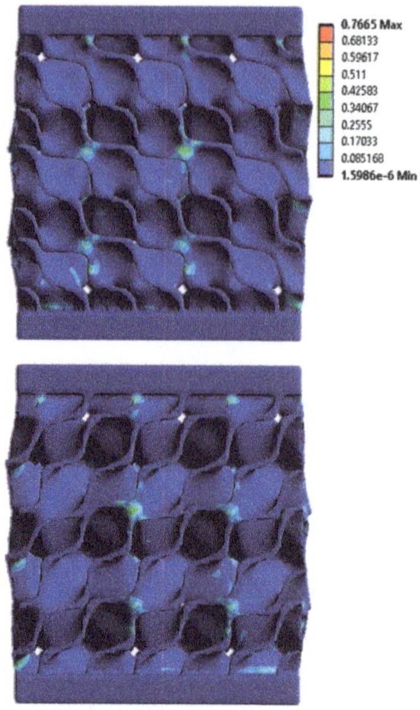	
S10 (R10)	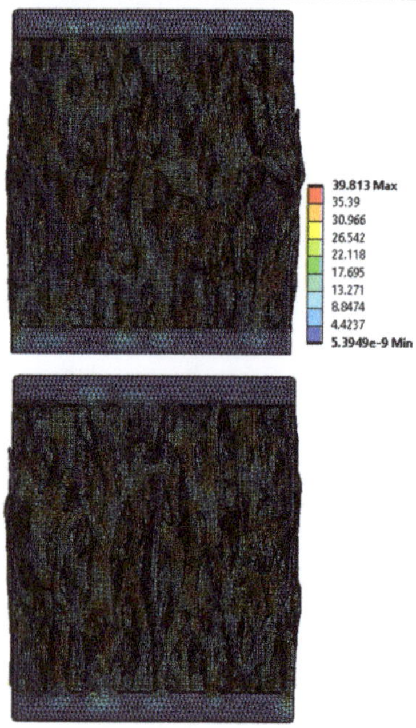	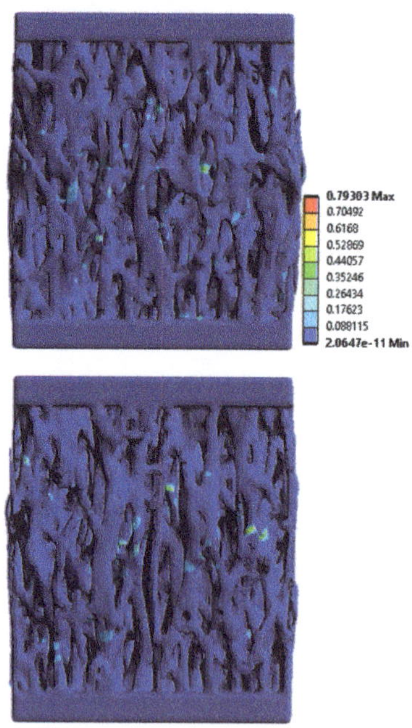	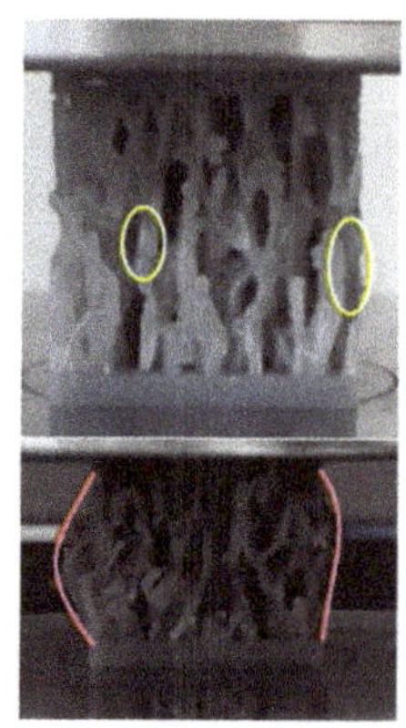

The distribution of equivalent stresses, in the case of specimen S1, showed high uniformity (column 2, row 1 notated as C2-R1) and together with the equivalent total strain (C3-R1), localized especially in the central area of the sample, leads to a deformation mechanism according to the one indicated in the figure in C4-R1.

Similarly, specimen S2 exhibited maximum stress values in the central layer of three-element cells (C2-R2), leading to maximum deformation in this area. This behavior is visible in the equivalent strain field (C3-R2) and the failure mode captured experimentally in C4-R2. For the S3 topology, the homogeneous stress distribution throughout the volume (C2-R3) indicates a stable failure behavior, as is visible in C4-R3. This, however, is not as easy to anticipate by analyzing only the equivalent strain field in C3-R3. Next, S4 faced convergence difficulties due to the thin walls generated on its sides and the inability to mesh them with a sufficient number of elements across their thickness. To address this, the deformation was limited to 1 mm, which only captures the elastic behavior of the specimen and the onset of yielding. Even at low deformation values, the regions with maximum equivalent stresses (C2-R4) coincide with the areas where deformation was first observed during experimental testing (C4-R4). Similarly, the equivalent total strain in C3-R4 indicates the critical regions between the layers, where yielding led to layer self-contacts, as highlighted in C4-R4. For specimen S5, the failure mechanism is easier to anticipate, with maximum equivalent stresses concentrated in the upper region (C2-R5) consistent with the deformation mode presented in C3-R5. Specimen S6 demonstrated the best uniformity, both in terms of equivalent von Mises stresses (C2-R6) and equivalent total strain (C3-R6). In this case, it was difficult to estimate the failure mode identified in C4-R6, with higher strain values in the lower zone of the specimen. For S7 the figures in C2-R7 show a leftward shift of the equivalent von Mises stresses, together with higher strain values in those zones suggested by C3-R7. Both these results are consistent with way the structure loses its equilibrium, identified experimentally in C3-R7, generated by buckling along the middle of the vertical walls of the RVE. As expected, specimen S8 presents the highest strains in the thin areas generated along the contour, which fractured at a specific strain of 20%. The distribution of equivalent von Mises stresses (C2-R8) and the most deformed areas (C3-R8) corresponded to empirical observations of 45° deformation angles on the outer walls (C4-R8). Topology S9 experienced the highest values of equivalent von Mises stresses (C2-R9) and equivalent total strain (C3-R9) in the areas between the inclined layers that make up the geometry, resulting in a folding failure mechanism where cells lay one on top of the other, as shown in C4-R9. Finally, specimen S10 also encountered convergence difficulties due to its stochastic nature, limiting the simulation to a 1 mm deformation. For this geometry, identifying a failure mechanism cannot be a realistic resolution, as the maximum equivalent stresses and total strains exhibited unpredictable behavior. Consequently, a direct comparison between experimental and FEA results was not feasible for this sample.

3.5. *Finite Element Analysis in Ansys Explicit Dynamics*

Next, the numerical simulation was conducted using the Ansys Explicit Dynamics module to evaluate the deformation mechanism across the entire range of experimentally imposed deformations. This type of FEA can handle cases of significant shape changes or lack of linearity of the material more efficiently.

Table 3 shows the variations in compressive force as a function of deformation, obtained experimentally and via FEA in both the *Static Structural* and *Explicit Dynamics* modules for the S1 and S8 topologies. The gyroid, S1, was selected as a reference due to its significance in relevant literature, while topology S8 was validated for demonstrating superior yield strength compared to the gyroid. Considering the constant input parameters of the two simulations, the overlap on the first 2.5 mm is evident. The slight difference between the two FEA curves (yellow and green) may be due to two factors. First, the analysis performed in the *Explicit Dynamics* module has a larger deformation increment, which can lead to less accurate approximations of the curve between successive points. Second, the definition of the *Multilinear* material is limited to a maximum of 10 points.

Table 3. Visualization of FEA results and experimental compressive response for samples S1 and S8.

Sample (C1)	Von Mises Equivalent Stress Distribution [MPa] (C2)	Comparison Between the Force–Deformation Plots and Those Obtained by Finite Element Analysis (C3)
S1 (R1)	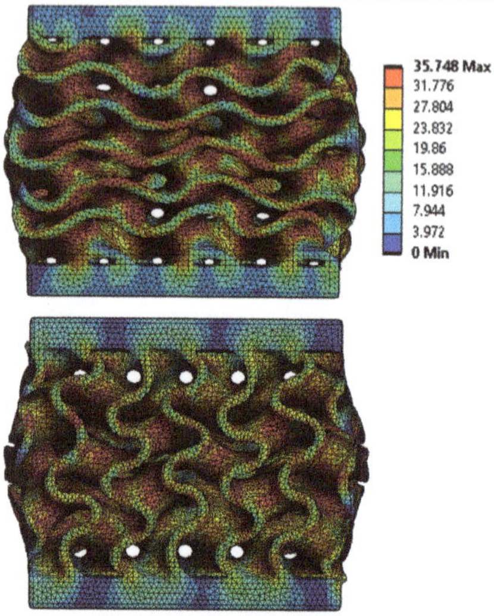	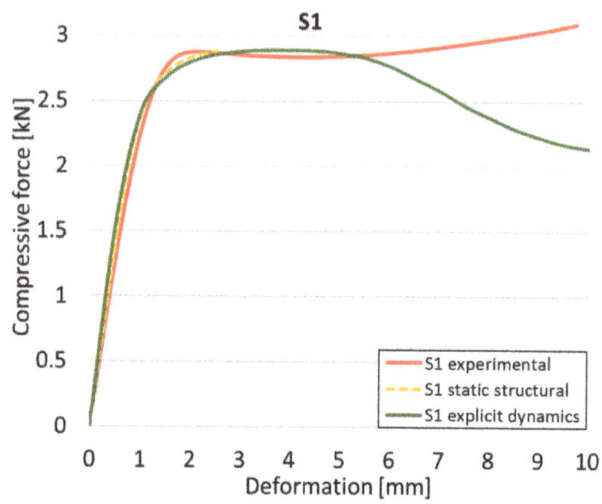
S8 (R2)	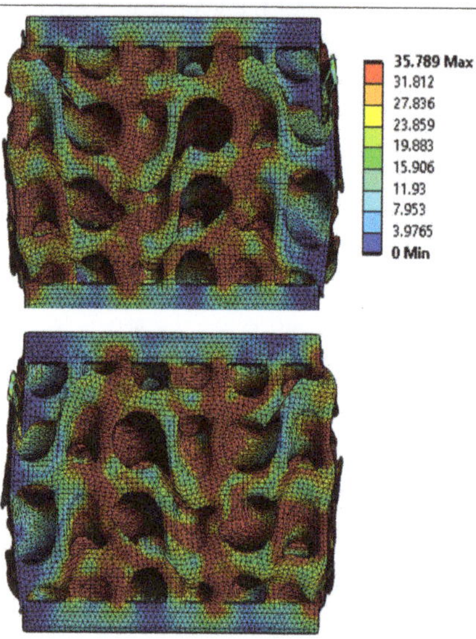	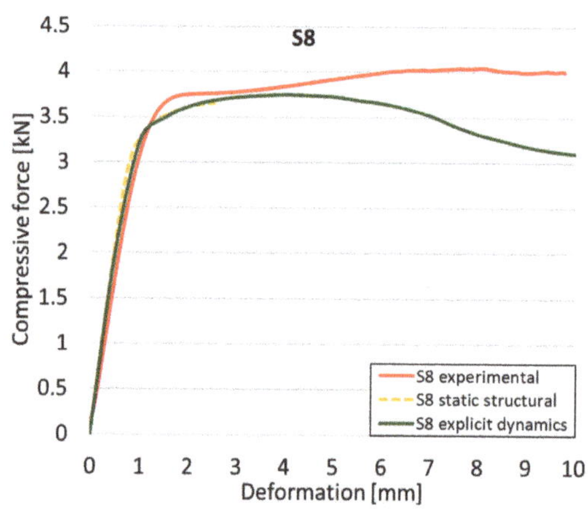

The proposed model does not adequately capture the non-linearity caused by the hardening phenomenon in the samples, as indicated by the reduced loading force at high strain values in both green curves. The figures in column C2 display the equivalent von Mises stress fields at a deformation of 10 mm. The general failure mechanism is similar to the one identified experimentally and presented in Table 2, column C4.

4. Discussion and Conclusions

After conducting a sensitivity analysis of the geometric model, it was determined that a maximum size of 0.3 mm for the regular tetrahedron elements offers an optimal balance between result accuracy and computational efforts. While this value is in line with similar sensitivity analyses from relevant literature [49,50], the analysis was specifically conducted to validate the mesh size for the proposed TPMS structures, given their intricate geometry and different dimensions. This corresponds to at least three elements along the thickness of the walls of the cellular solids. However, employing a uniform meshing strategy throughout the entire part may prove to be insufficient in the case of geometries with walls whose thickness is not constant or in areas with very low thicknesses, where at least two rows of elements cannot be generated. This, in turn, results in convergence problems or accuracy reduction, as observed in the cases of topologies S4 and S10.

An analysis was conducted on the implications of using implicit modeling in FEA. For the same gyroid topology, S1, numerical analyses were carried out using geometric models created through both conventional CAD methods and implicit modeling. The dimensional deviations between the models exceeded the imposed tolerance value of 0.1 mm, with maximum values of 0.17 mm. However, the resulting variation of compressive force as a function of deformation showed a very good overlap between the two models. A disadvantage of using geometries obtained through implicit modeling is the significantly longer time required to establish the faceted geometry, transform the geometry into a solid, perform the meshing, and solve the model. This can be up to 4 times longer than in the case of a conventional approach, which makes the finer meshing of solids more challenging due to the rapid increase in computational effort.

Four material models were proposed to identify which best approximates the behavior of the selected photopolymer resin. Data obtained from tensile tests were used to empirically define two isotropic material models: *Bilinear* and *Multilinear*. Following this, an iterative method was employed to configure two additional material models: *Powerlaw* and the *Three-Network Model*. The parameters defining their equations were configured so that the characteristic curves of the materials overlapped as accurately as possible over the average true stress–strain curve obtained through tensile tests. FEAs were conducted for all four material models, and the error between the FEA results and the experimental data determined through three different compression tests was assessed. Deviations of less than 6% were found, with the *Multilinear* model achieving the best accuracy at only 2.54% deviation. This is in line with findings in relevant literature for isotropic hardening models such as the analyses performed by Zheng et al. in [10], Naghavi et al. in [30], or Krešic et al. in [51]. Also, as suggested by these studies, the numerical results for compressive force after yielding tends to have lower values than the experimental data.

Performing quasi-static numerical analyses that include high strain values on complex three-dimensional geometries can prove difficulties due to the contact interactions within the samples. Increasing the tolerance of the walls penetrating each other can increase the convergence rate but will also lead to reduced accuracy. In this sense, static tests were performed for a deformation of only 2.5 mm, a value chosen so that there would be no contacts that would lead to convergence difficulties. Even in this case, due to the periodic nature that determined thin walls on specimen S4 and the stochastic nature of sample S10, the deformation value was reduced to 1 mm to achieve convergence, even though this limited the range of phenomena captured.

By corroborating the equivalent von Mises stresses and equivalent total strain fields obtained for each sample, conclusions were drawn regarding the general failure modes of the samples. In most instances, the FEA results indicated a deformation mechanism similar to what was observed in experimental testing. However, sample S6 was an exception, as the substantial deformation occurring in the lower part of this specimen was not captured in the numerical simulation. This discrepancy could be due to the limited strain value imposed in the simulation, or because the real failure mode was influenced by manufacturing defects which were not considered in the FEA.

Using implicit models in FEA is a viable approach when traditional CAD models are unavailable due to complex designs. The numerical simulation method proposed in this paper automatically generates or optimizes innovative topologies through implicit modeling, allowing for the validation of their properties before manufacturing takes place. This process will be the focus of a future study, which will involve the automatic generation and evaluation of structures using a Python–Ansys interface.

Author Contributions: Conceptualization, A.V. and D.M.C.; methodology, A.V. and A.I.I.; software, A.V. and I.C.C.; validation, D.M.C., Ş.S. and A.I.I.; formal analysis, D.M.C.; investigation, A.V. and I.C.C.; resources, Ş.S.; data curation, A.I.I. and D.A.A.; writing—original draft preparation, A.V.; writing—review and editing, D.M.C., Ş.S., A.I.I. and D.A.A.; visualization, A.V.; supervision, D.M.C. and Ş.S.; project administration, Ş.S.; funding acquisition, D.M.C. All authors have read and agreed to the published version of the manuscript.

Funding: Alexandru Vasile acknowledges the grant awarded by the Ministry of Education from Romania for completing his PhD studies under contract no. 06.87/05.10.2020.

Institutional Review Board Statement: Not applicable.

Informed Consent Statement: Not applicable.

Data Availability Statement: The original contributions presented in the study are included in the article, further inquiries can be directed to the corresponding author.

Conflicts of Interest: The authors declare no conflicts of interest.

References

1. Lu, C.; Hsieh, M.; Huang, Z.; Zhang, C.; Lin, Y.; Shen, Q.; Chen, F.; Zhang, L. Architectural Design and Additive Manufacturing of Mechanical Metamaterials: A Review. *Engineering* **2022**, *17*, 44–63. [CrossRef]
2. Kladovasilakis, N.; Tsongas, K.; Karalekas, D.; Tzetzis, D. Architected Materials for Additive Manufacturing: A Comprehensive Review. *Materials* **2022**, *15*, 5919. [CrossRef] [PubMed]
3. Feng, Y.; Qiu, H.; Gao, Y.; Zheng, H.; Tan, J. Creative Design for Sandwich Structures: A Review. *Int. J. Adv. Robot. Syst.* **2020**, *17*. [CrossRef]
4. Gado, M.G.; Al-Ketan, O.; Aziz, M.; Al-Rub, R.A.; Ookawara, S. Triply Periodic Minimal Surface Structures: Design, Fabrication, 3D Printing Techniques, State-of-the-Art Studies, and Prospective Thermal Applications for Efficient Energy Utilization. *Energy Technol.* **2024**, *12*, 2301287. [CrossRef]
5. Feng, J.; Fu, J.; Yao, X.; He, Y. Triply Periodic Minimal Surface (TPMS) Porous Structures: From Multi-Scale Design, Precise Additive Manufacturing to Multidisciplinary Applications. *Int. J. Extrem. Manuf.* **2022**, *4*, 022001. [CrossRef]
6. Yu, X.; Zhou, J.; Liang, H.; Jiang, Z.; Wu, L. Mechanical Metamaterials Associated with Stiffness, Rigidity and Compressibility: A Brief Review. *Prog. Mater. Sci.* **2018**, *94*, 114–173. [CrossRef]
7. Hassan, I.M.; Enab, T.A.; Fouda, N.; Eldesouky, I. Design, Fabrication, and Evaluation of Functionally Graded Triply Periodic Minimal Surface Structures Fabricated by 3D Printing. *J. Braz. Soc. Mech. Sci. Eng.* **2023**, *45*, 66. [CrossRef]
8. Wakjira, Y.; Cioni, A.; Lemu, H.G. Current Status of the Application of Additive-Manufactured TPMS Structure in Bone Tissue Engineering. *Prog. Addit. Manuf.* **2024**. [CrossRef]
9. Echeta, I.; Dutton, B.; Leach, R.K.; Piano, S. Finite Element Modelling of Defects in Additively Manufactured Strut-Based Lattice Structures. *Addit. Manuf.* **2021**, *47*, 102301. [CrossRef]
10. Zhang, L.; Feih, S.; Daynes, S.; Chang, S.; Wang, M.Y.; Wei, J.; Lu, W.F. Energy Absorption Characteristics of Metallic Triply Periodic Minimal Surface Sheet Structures under Compressive Loading. *Addit. Manuf.* **2018**, *23*, 505–515. [CrossRef]
11. Qiu, N.; Wan, Y.; Shen, Y.; Fang, J. Experimental and Numerical Studies on Mechanical Properties of TPMS Structures. *Int. J. Mech. Sci.* **2024**, *261*, 108657. [CrossRef]
12. Li, F.; Gan, J.; Zhang, L.; Tan, H.; Li, E.; Li, B. Enhancing Impact Resistance of Hybrid Structures Designed with Triply Periodic Minimal Surfaces. *Compos. Sci. Technol.* **2024**, *245*, 110365. [CrossRef]

13. Li, X.; Xiao, L.; Song, W. Compressive Behavior of Selective Laser Melting Printed Gyroid Structures under Dynamic Loading. *Addit. Manuf.* **2021**, *46*, 102054. [CrossRef]
14. Yang, E.; Leary, M.; Lozanovski, B.; Downing, D.; Mazur, M.; Sarker, A.; Khorasani, A.M.; Jones, A.; Maconachie, T.; Bateman, S.; et al. Effect of Geometry on the Mechanical Properties of Ti-6Al-4V Gyroid Structures Fabricated via SLM: A Numerical Study. *Mater. Des.* **2019**, *184*, 108165. [CrossRef]
15. Teng, F.; Sun, Y.; Guo, S.; Gao, B.; Yu, G. Topological and Mechanical Properties of Different Lattice Structures Based on Additive Manufacturing. *Micromachines* **2022**, *13*, 1017. [CrossRef]
16. Afshar, M.; Anaraki, A.P.; Montazerian, H.; Kadkhodapour, J. Additive Manufacturing and Mechanical Characterization of Graded Porosity Scaffolds Designed Based on Triply Periodic Minimal Surface Architectures. *J. Mech. Behav. Biomed. Mater.* **2016**, *62*, 481–494. [CrossRef] [PubMed]
17. Zheng, X.; Fu, Z.; Du, K.; Wang, C.; Yi, Y. Minimal Surface Designs for Porous Materials: From Microstructures to Mechanical Properties. *J. Mater. Sci.* **2018**, *53*, 10194–10208. [CrossRef]
18. Yang, L.; Mertens, R.; Ferrucci, M.; Yan, C.; Shi, Y.; Yang, S. Continuous Graded Gyroid Cellular Structures Fabricated by Selective Laser Melting: Design, Manufacturing and Mechanical Properties. *Mater. Des.* **2019**, *162*, 394–404. [CrossRef]
19. Feng, G.; Li, S.; Xiao, L.; Song, W. Mechanical Properties and Deformation Behavior of Functionally Graded TPMS Structures under Static and Dynamic Loading. *Int. J. Impact Eng.* **2023**, *176*, 104554. [CrossRef]
20. Zhang, J.; Xie, S.; Li, T.; Liu, Z.; Zheng, S.; Zhou, H. A Study of Multi-Stage Energy Absorption Characteristics of Hybrid Sheet TPMS Lattices. *Thin-Walled Struct.* **2023**, *190*, 110989. [CrossRef]
21. Ejeh, C.J.; Barsoum, I.; Abou-Ali, A.M.; Abu Al-Rub, R.K. Combining Multiple Lattice-Topology Functional Grading Strategies for Enhancing the Dynamic Compressive Behavior of TPMS-Based Metamaterials. *J. Mater. Res. Technol.* **2023**, *27*, 6076–6093. [CrossRef]
22. Zhang, H.; Zhao, J.; Niu, Q.; Xu, C.; Huo, R. A Study of Energy Absorption Properties of Heteromorphic TPMS and Multi-Morphology TPMS under Quasi-Static Compression. *Thin-Walled Struct.* **2024**, *205*, 112519. [CrossRef]
23. Alawwa, F.; Barsoum, I.; Abu Al-Rub, R.K. Modeling, Testing, and Optimization of Novel Lattice Structures for Enhanced Mechanical Performance. *Mech. Adv. Mater. Struct.* **2023**, *31*, 8691–8714. [CrossRef]
24. Guo, A.; Li, S.; Wang, S.; Zhai, Z.; Qu, P.; Guo, S.; Kong, H.; Tang, R.; Wang, L.; Meng, W.; et al. Novel Triply Periodic Minimal Surfaces Sandwich Structures: Mechanical Performance and Failure Analysis. *Polym. Compos.* **2024**, *45*, 11908–11924. [CrossRef]
25. Luo, Z.; Tang, Q.; Feng, Q.; Ma, S.; Song, J.; Setchi, R.; Guo, F.; Zhang, Y. Finite Element Analysis of the Mechanical Properties of Sheet- and Skeleton-Gyroid Ti6Al4V Structures Produced by Laser Powder Bed Fusion. *Thin-Walled Struct.* **2023**, *192*, 111098. [CrossRef]
26. Huang, X.; Tang, H.; Wang, L. Effect of Residual Stress on Mechanical Properties of Triply Periodic Minimal Surface Lattice Structures in Additive Manufacturing. *Comput. Mater. Sci.* **2024**, *245*, 113318. [CrossRef]
27. Verma, R.; Kumar, J.; Singh, N.K.; Rai, S.K.; Saxena, K.K.; Xu, J. Design and Analysis of Biomedical Scaffolds Using TPMS-Based Porous Structures Inspired from Additive Manufacturing. *Coatings* **2022**, *12*, 839. [CrossRef]
28. Lin, Z.H.; Pan, J.H.; Li, H.Y. Mechanical Strength of Triply Periodic Minimal Surface Lattices Subjected to Three-Point Bending. *Polymers* **2022**, *14*, 2885. [CrossRef]
29. Yadav, P.; Kushwaha, A.; Shukla, M. Analysis of Compression Behavior of TPMS and Strut Type Lattice Structures Using Finite Element Method. *NanoWorld J.* **2023**, *9*, S687–S692. [CrossRef]
30. Naghavi, S.A.; Tamaddon, M.; Marghoub, A.; Wang, K.; Babamiri, B.B.; Hazeli, K.; Xu, W.; Lu, X.; Sun, C.; Wang, L.; et al. Mechanical Characterisation and Numerical Modelling of TPMS-Based Gyroid and Diamond Ti6Al4V Scaffolds for Bone Implants: An Integrated Approach for Translational Consideration. *Bioengineering* **2022**, *9*, 504. [CrossRef] [PubMed]
31. Abueidda, D.W.; Elhebeary, M.; Shiang, C.S.; Pang, S.; Abu Al-Rub, R.K.; Jasiuk, I.M. Mechanical Properties of 3D Printed Polymeric Gyroid Cellular Structures: Experimental and Finite Element Study. *Mater. Des.* **2019**, *165*, 107597. [CrossRef]
32. Wang, Y.; Liu, F.; Zhang, X.; Zhang, K.; Wang, X.; Gan, D.; Yang, B. Cell-Size Graded Sandwich Enhances Additive Manufacturing Fidelity and Energy Absorption. *Int. J. Mech. Sci.* **2021**, *211*, 106798. [CrossRef]
33. Wan, M.; Hu, D.; Zhang, H.; Zhang, Z. Energy Absorption Characteristics of TPMS-Filled Square Tubes under Quasi-Static Axial Crushing. *Thin-Walled Struct.* **2024**, *199*, 111811. [CrossRef]
34. Vasile, A.; Constantinescu, D.M.; Coropețchi, I.C.; Sorohan, Ș.; Apostol, D.A. Definition, Fabrication, and Compression Testing of Sandwich Structures with Novel TPMS-Based Cores. *Materials* **2024**, *17*, 5150. [CrossRef] [PubMed]
35. Geuzaine, C.; Remacle, J.F. Gmsh: A 3-D Finite Element Mesh Generator with Built-in Pre- and Post-Processing Facilities. *Int. J. Numer. Methods Eng.* **2009**, *79*, 1309–1331. [CrossRef]
36. Nguyen, C.H.P.; Kim, Y.; Do, Q.T.; Choi, Y. Implicit-Based Computer-Aided Design for Additively Manufactured Functionally Graded Cellular Structures. *J. Comput. Des. Eng.* **2021**, *8*, 813–823. [CrossRef]
37. Chougrani, L.; Pernot, J.P.; Véron, P.; Abed, S. Lattice Structure Lightweight Triangulation for Additive Manufacturing. *Comput. Aided Des.* **2017**, *90*, 95–104. [CrossRef]

38. *ASTM D638-14*; Standard Test Method for Tensile Properties of Plastics. ASTM International: West Conshohocken, PA, USA, 2016
39. Vasile, A.; Constantinescu, D.M.; Coropețchi, I.C.; Sorohan, Ș.; Apostol, D.A. Assessment of the Additive Fabrication Quality of Sandwich Structures with Novel Triply Periodic Minimal Surface Cores. *Solids* **2024**, *5*, 665–680. [CrossRef]
40. Decker, T.; Kedziora, S.; Museyibov, E. Comparison of Strength Properties of Common Powder Bed Fusion and Stereolithography Materials. *Key Eng. Mater.* **2023**, *969*, 11–20. [CrossRef]
41. Gao, S.; Liu, W.; Zhang, L.; Gain, A.K. A New Polymer-Based Mechanical Metamaterial with Tailorable Large Negative Poisson's Ratios. *Polymers* **2020**, *12*, 1492. [CrossRef] [PubMed]
42. *ANSYS Mechanical APDL Plasticity*; Rate-Independent Plasticity; ANSYS, Inc.: Canonsburg, PA, USA, 2024; Volume R2.
43. Volgin, O.; Shishkovsky, I. Material Modelling of FDM Printed PLA Part. *Eng. Solid Mech.* **2021**, *9*, 153–160. [CrossRef]
44. Kumar, S.; Singh, S.S.; Rozycki, P. Numerical Simulation of Strain-Softening Behavior of Glass-Filled Polymer Composites: Comparison of Two-Dimensional and Three-Dimensional Analyses Using Arruda-Boyce and Three-Network Viscoplastic Models. *Mech. Mater.* **2022**, *175*, 104481. [CrossRef]
45. Chai, B.X.; Eisenbart, B.; Nikzad, M.; Fox, B.; Wang, Y.; Bwar, K.H.; Zhang, K. Review of Approaches to Minimise the Cost of Simulation-Based Optimisation for Liquid Composite Moulding Processes. *Materials* **2023**, *16*, 7580. [CrossRef]
46. *ANSYS Mechanical APDL Element Reference*; Ansys, Inc.: Canonsburg, PA, USA, 2022; pp. 861–868.
47. *ANSYS Mechanical APDL Contact Technology Guide*; Ansys, Inc.: Canonsburg, PA, USA, 2009; pp. 21–27.
48. Brinson, H.F.; Brinson, L.C. *Polymer Engineering Science and Viscoelasticity: An Introduction*; Springer: New York, NY, USA, 2008; ISBN 9780387738604.
49. Wang, Z.; Xu, M.; Du, J.; Jin, Y. Experimental and Numerical Investigation of Polymer-Based 3D-Printed Lattice Structures with Largely Tunable Mechanical Properties Based on Triply Periodic Minimal Surface. *Polymers* **2024**, *16*, 711. [CrossRef] [PubMed]
50. Samson, S.; Tran, P.; Marzocca, P. Design and Modelling of Porous Gyroid Heatsinks: Influences of Cell Size, Porosity and Material Variation. *Appl. Therm. Eng.* **2023**, *235*, 121296. [CrossRef]
51. Krešic, I.K.; Kaljun, J.; Rašovic, N. Controlling the Mechanical Response of Stochastic Lattice Structures Utilizing a Design Model Based on Predefined Topologic and Geometric Routines. *Appl. Sci.* **2024**, *14*, 6048. [CrossRef]

materials

MDPI

Article

Evaluating the Role of Unit Cell Multiplicity in the Acoustic Response of Phononic Crystals Using Laser-Plasma Sound Sources

Emmanouil Kaniolakis Kaloudis [1,2], Konstantinos Kaleris [1,2,*], Nikos Aravantinos-Zafiris [3], Michael Sigalas [4], Dionysios T. G. Katerelos [5], Vasilis Dimitriou [1,2], Makis Bakarezos [1,2], Michael Tatarakis [1,6] and Nektarios A. Papadogiannis [1,2]

1 Institute of Plasma Physics and Lasers (IPPL), Hellenic Mediterranean University, Tria Monastiria, GR-74100 Rethymnon, Greece; mkaniolakis@hmu.gr (E.K.K.); dimvasi@hmu.gr (V.D.); bakarezos@hmu.gr (M.B.); mictat@hmu.gr (M.T.); npapadogiannis@hmu.gr (N.A.P.)
2 Physical Acoustics and Optoacoustics Laboratory, Department of Music Technology and Acoustics, Hellenic Mediterranean University, GR-74100 Rethymnon, Greece
3 Department of Environment, Ionian University, GR-29100 Zakynthos, Greece; naravadinos@ionio.gr
4 Department of Materials Science, University of Patras, GR-26500 Patras, Greece; sigalas@upatras.gr
5 Department of Audio & Visual Arts, Ionian University, GR-49100 Corfu, Greece; dkaterelos@ionio.gr
6 Department of Electronic Engineering, Hellenic Mediterranean University, GR-73133 Chania, Greece
* Correspondence: kkaleris@hmu.gr

Abstract: Acoustic metamaterials and phononic crystals are progressively consolidating as an important technology that is expected to significantly impact the science and industry of acoustics in the coming years. In this work, the impact of unit cell multiplicity on the spectral features of the acoustic response of phononic crystals is systematically studied using the recently demonstrated laser-plasma sound source characterization method. Specifically, by exploiting the advantages of this method, the impact of the number of repeated unit cells on the depth of the phononic band gaps and the passband spectral features across the entire audible range is demonstrated. These experimental findings are supported by specially developed computational simulations accounting for the precise structural characteristics of the studied phononic crystals and are analysed to provide a phenomenological understanding of the underlying physical mechanism. It is shown that by increasing the unit cell multiplicity, the bandgaps deepen and the number of resonant peaks in the crystal transmission zones increases. The resonant mode shapes are computationally investigated and interpreted in terms of spherical harmonics. This study highlights the tunability and design flexibility of acoustic components using phononic crystals, opening new paths towards applications in the fields of sound control and noise insulation.

Keywords: acoustic metamaterials; phononic crystals; laser-plasma sound sources; acoustic characterization

1. Introduction

Acoustic metamaterials (AMs) and phononic crystals (PCs) are engineered materials that manipulate sound in ways that are not possible with conventional acoustic materials. Particularly, AMs exploit structures shorter than the wavelength of the sound they control to achieve unconventional behaviour like negative refraction and acoustic cloaking [1,2]. Conversely, PCs are artificial periodic arrangements of multiple components (unit cells) that affect acoustic wave propagation due to Bragg scattering. Each unit cell features an internal structure with regions of differing acoustic properties, such as acoustic impedance and/or sound speed [3–5]. Interaction between propagating acoustic waves and PCs results in a characteristic frequency response consisting of forbidden and permitted spectral regions, named phononic band gaps (stopbands) and transmission bands (passbands), respectively [6]. The acoustic spectral profile of PC is related to Bragg scattering [7–10], which takes place when the wavelength of the propagating acoustic wave (AW) is comparable to the features and periodicity of the structure [11]. Specifically, when the AWs propagate through the structure, they encounter regions of different acoustic impedances, forming interfaces upon which the AWs scatter. Superposition of the scattered waves from multiple layers gives rise to interference patterns. When the Bragg condition for constructive interference is satisfied, wave propagation is allowed, and the passband regions are formed. Conversely, for the frequencies satisfying the destructive interference condition, acoustic transmission is prohibited, leading to the emergence of spectral bandgaps.

Importantly, the passbands and stopbands of PCs can be precisely engineered by designing the geometrical and structural characteristics of the unit cells. Their capability to tailor their acoustic response makes PCs ideal for various applications. Particularly, PCs are utilized in noise insulation [12–14], vibration isolation [15,16], sound attenuation [17] and soundproofing [18], where the targeted attenuation of desired frequencies is achieved by exploiting phononic crystal bandgaps. Additionally, control of the passband features facilitates the design of acoustic waveguides [19–24] and acoustic filters [5,11,25–27], enabling the selective transmission of AWs by amplifying the transmission in specific frequencies. Finally, apart from precisely tailored acoustic filtering [21], specially designed PCs can bend, focus or redirect the propagating sound waves [28–30], rendering them useful in applications such as acoustic cloaking and directional sound generation [31,32].

The complexity of acoustic wave-structure interaction in PCs often renders the comprehension and analytical description of their behavior difficult. Currently, the design of new PCs is commonly done via complex numerical simulations and is often based on the experience of the designer rather than on a systematic approach. Moreover, experimental evaluation of developed structures on the level of a single or a few unit cells has proved difficult with conventional methods. The recently demonstrated Laser-Plasma Sound Source (LPSS) method [33] allows for evaluation of acoustic features of the PC transmission spectrum, including band gap depth and resonant features, with unprecedented accuracy. This work exploits the LPSS method to identify the impact of the number of unit cells on the transmission spectra of PCs across the entire audible range. Particularly, we experimentally evaluate the transmission spectra of linear PCs with different lengths, developed via repetition of the same unit cell. LPSSs are highly effective in characterizing PCs due to their massless and point-like geometry, rapid pressure profile, broad frequency spectrum and high acoustic energy [33]. Particularly, we focus on the impact of the number of unit cells on the following: (a) the depth of the bandgaps; and (b) the profile of the resonant features in the passbands of the acoustic transmission spectrum within the entire audible range. Our findings reveal a clear connection between the number of unit cells and the depth of the gaps. Specifically, the depth increases with the increasing number of unit cells. Furthermore, an increase in the cell multiplicity leads to an increase in the number of resonant peaks within the second and third passbands. The experimental observations are validated by specially developed computational simulations, showing excellent agreement. This work reveals the exact role of the structural multiplicity in the shaping and control of the spectral features of PCs. More specifically, it provides an experimental methodology for the determination of the number of unit cells required to achieve a specific bandgap depth, especially with respect to the behavior of a structure with infinite length. It also demonstrates the impact of cell multiplicity on the resonant features of the passbands, which can be directly related to the spectral flatness of the transmission regions. To the best of our knowledge, there are no publications that provide computationally validated experimental evidence on these aspects. This is potentially due to the prior absence of sufficiently accurate experimental methods for their investigation, a shortcoming that is here overcome by the LPSS method. In this respect, the study contributes to the systematization of the design and optimization process of PCs. More generally, it can lead to efficient PC designs with tailored control over AW transmission, which can be useful in various scientific and industrial applications, especially in room acoustics and noise control.

2. Materials and Methods

The phononic crystals used in this study were developed via the repetition of a cylindrical unit cell consisting of a spherical air cavity surrounded by six air conduits. The samples were fabricated in house using polylactic acid (PLA) (XYZprinting), with the fused filament technique on an 3D printer (XYZprinting da Vinci Super, New Taipei Taiwan). The unit cells had a lattice constant $a = 29$ mm, with a sphere radius of $R = 0.395a$, a conduit radius of $r = 0.095a$ and an infill of 100%. The infill value represents the porosity of the crystal, with an infill of 100% corresponding to a porosity of 0%. Four structures were fabricated with the repetition of 2, 3, 4 and 5 unit cells (Figure 1). This structure has already been shown, both experimentally and numerically, that provides wide and complete acoustic band gaps [9,33].

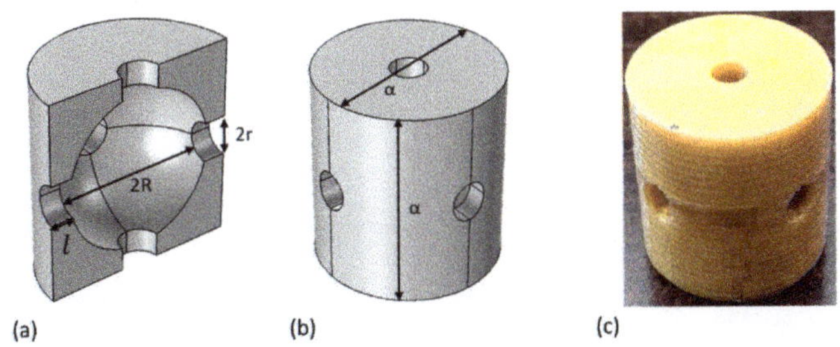

Figure 1. Internal (**a**) and external (**b**) design of the unit cell, where $a = 29$ mm is the lattice constant, $R = 0.395a$ is the sphere radious, $r = 0.095a$ is the conduit radious and $l = 0.105\alpha$ is the conduit length. (**c**) Image of a 3D-printed unit cell made of PLA.

Evaluation of the impact of the unit cell multiplicity on the transmission spectrum of the PCs was carried out by the use of laser plasma sound sources. LPSSs are generated in ambient air by the thermoelastic reaction of a localized air volume following Laser-Induced Breakdown (LIB). High-intensity, short or ultrashort laser pulses ionize the air particles, generating high-temperature free electrons via multiphoton ionization and inverse Bremsstrahlung. These hot electrons thermalize the heavier ions via various physical processes, e.g., recombinations, causing the rapid thermal expansion of the air volume and the formation of a characteristic acoustic N-pulse [33]. The laser generated acoustic N-pulses exhibit a first order high-pass profile at the lower end of the acoustic spectrum. For excitation from nanosecond laser pulses, this first order profile spans from the infrasounds (<20 Hz) to the high end of the audible spectrum (~20 kHz) or even the near ultrasounds (~60 kHz). Peak pressure levels can exceed 130 dB, depending on the optical energy deposited by the laser pulses in the excitation volume. Extensive studies of the N-pulses and their relation to the laser pulse characteristics and parameters can be found in [34–38].

In our experiments, LPSS formation was done by a 6 nanosecond Nd: YAG pulsed laser system (Quantel Brilliant B) with 1064 nm central wavelength and up to 10 Hz repetition rate. Here, a repetition rate of 5 Hz was used that allows for a 200 ms time interval between consecutive pulses. The laser beam was focused with a 7.5 cm bi-convex lens in the ambient air. The laser pulse energy of 5.5 mJ ensured that the laser-plasma source was perfectly spherical and hence omnidirectional, as evaluated by fast shadowgraphy. The LPSSs were positioned ~2 mm from the input hole of the structure. The acoustic measurement system was based on a broad frequency range (90 kHz) and high dynamic range (35 dB(A)–160 dB) microphone (G.R.A.S 46BE and power module 12AK). A sound card (RME Fireface 802) with a 192 kHz sampling rate and 24-bit resolution was used for the sampling of the acoustic signals. The microphone was placed at a distance of ~3 mm from the output hole, so that it was sufficiently close to the structure but without blocking the output hole. The experimental set up is shown in Figure 2. The acoustic signals were recorded with the Audacity software (version 3.7.0) [39]. Each measurement contained a train of 120 consecutive responses, corresponding to a total duration of 24 s. Noise reduction was achieved by taking the average waveform of the 120 acoustic responses. An analytical description of the signal processing methodology can be found in [33].

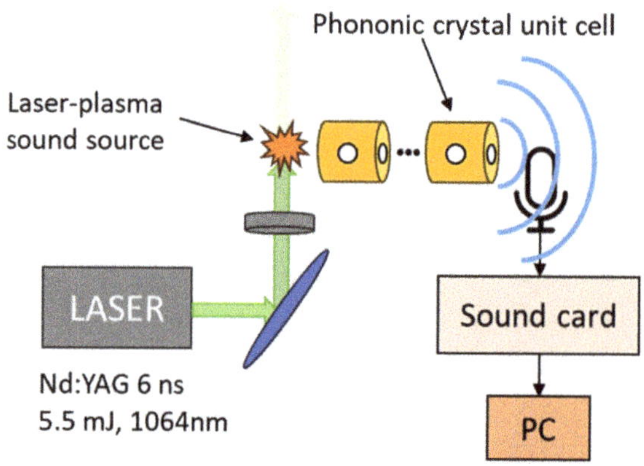

Figure 2. Schematic diagram of the experimental setup used for the acoustic evaluation of phononic crystals via laser-plasma sound source excitation.

Finally, numerical simulations regarding the PCs' acoustic response were performed using COMSOL Multiphysics® (version 5.5) and particularly the Acoustics Module and the Solid Mechanics interface. The Finite Element Method included in the commercial package was used to numerically solve the wave propagation following equation for sound:

$$\nabla\left(-\frac{1}{\rho_c}(\nabla p_t - q_d)\right) - \frac{k_{eq}^2 p_t}{\rho_c} = \frac{4\pi}{\rho_c} S\delta(x - x_0) \tag{1}$$

$$S = \omega e^{i\varphi}\sqrt{\frac{\rho_c u_{air} p_{rms}}{2\pi}} \tag{2}$$

where p_t is the total acoustic pressure, ρ_c is the density of ambient air, $k_{eq} = \frac{2\pi}{\omega}$ is the wavenumber and ω the angular frequency, q_d is the dipole domain source and $\delta(x - x_0)$ is the Dirac's delta function. Additionally, S is the amplitude of the monopole point source, p_{rms} is the reference pressure and u_{air} is the speed of sound in ambient air. The source spectrum exhibits a 1st order high pass profile ($S \sim \omega$) in order to emulate the LPSS spectrum within the frequency range of interest.

For the numerical calculations, the following values for the properties of the PLA were used: Poisson ratio $v = 0.35$, Young's Modulus $E = 3.53$ GPa and density $\rho = 1240$ kgm^{-3}. The acoustic-solid interaction frequency domain interface was used in the model for the coupling of the ambient air with the solid boundary. A monopole point source of flow type was used to represent the laser-plasma sound source within the audible frequency range. Perfectly Matched Layers (PML) were used at the boundaries of the computational domain to eliminate boundary reflections. The transmission spectra are calculated according to the following equation:

$$T = 20log_{10}\frac{p_T}{p_I} \quad (3)$$

where p_T and p_I represent the transmitted and incident pressure field on the two sides of the sample.

3. Results

Figure 3a,b show the measured and simulated spectral responses of the PCs to LPSS excitation in the frequency range from 1.5 to 20 kHz. The laser-plasma source was placed at a distance of ~2 mm from the input hole, resulting in acoustic wave propagation along the depth of the PC. The experimental results shown in Figure 3a clearly reveals two phononic bandgaps between 3.6 and 9.4 kHz and between 11 and 13.6 kHz, as well as three passbands with characteristic resonant peaks at approximately 2.5–3.5 kHz, 9.5–11 kHz and 13.5–19.5 kHz. As expected, the frequency ranges of both the phononic bandgaps and the passbands remained approximately the same, independently from the number of unit cells. However, the phononic bandgaps became deeper with increases in the number of unit cells, as the structure's behaviour tends towards that of the infinite structure.

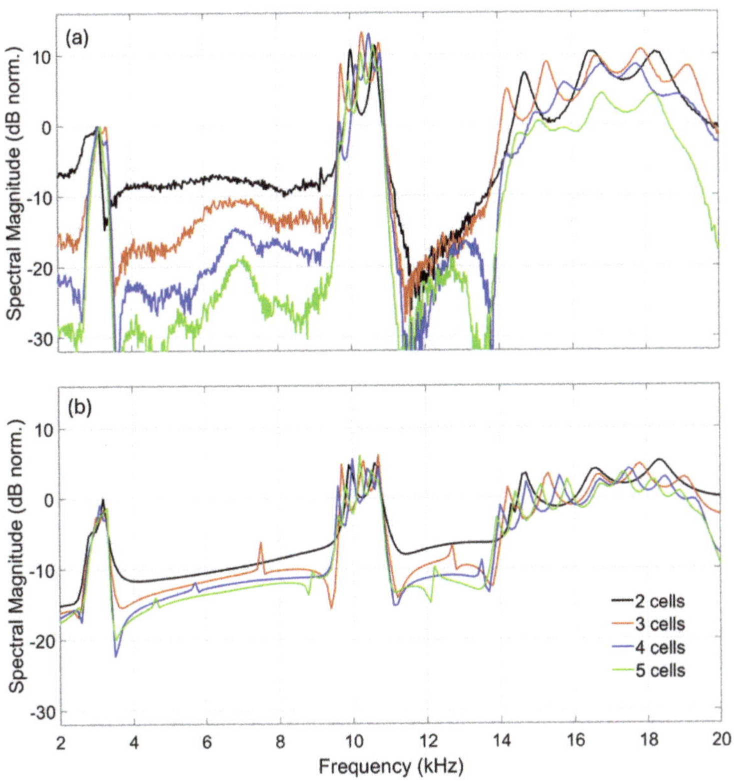

Figure 3. Experimentally measured (**a**) and simulated (**b**) acoustic transmission responses of phononic crystals consisting of 2–5 unit cells to laser-plasma sound source excitation.

To quantify the deepening of the band gap, the energy ratio E_R is calculated as follows:

$$E_R = E_{\text{bg1}}^{(n)} / E_{\text{bg1}}^{(2)} \tag{4}$$

$E_{\text{bg1}}^{(n)}$ is the energy in the 1st band gap of an n-cell structure and is given by the following:

$$E_{\text{bg1}}^{(n)} = \sum_{f_{\text{low}}}^{f_{\text{up}}} S^2(f) \tag{5}$$

where $f_{low} = 3850$ Hz, $f_{\text{up}} = 9050$ Hz denote the lower and upper frequencies of the band gap and $S^2(f)$ is the spectral energy obtained by fast Fourier transform (FFT) of the transmission signal $s_T(t)$, expressible as the following:

$$S^2(f) = (\text{FFT}\{s_T(t)\})^2 \tag{6}$$

In E_R, the energy of the n-cell structure is compared to that of the 2-cell structure $\left(E_{\text{bg1}}^{(2)}\right)$, which is considered to be the PC with the lowest cell multiplicity. The energy ratio taken by the experimental measurements can be seen in the scatter plot of Figure 4 (experimental data). One step further, E_R can be used for the estimation of the sufficient number of unit cells for the PC to approximately function as an infinite phononic crystal. For this purpose, we have adopted the criterion of band gap energy drop by a factor of $1/e$ (~36.7%), expressible as the following:

$$\frac{E_{\text{bg1}}^{(n)}}{E_{\text{bg1}}^{(2)}} \leq \frac{1}{e} \Rightarrow 10\log_{10}\left(\frac{E_{\text{bg1}}^{(n)}}{E_{\text{bg1}}^{(2)}}\right) \leq -4.34 \text{ dB} \tag{7}$$

where e is the base of the natural logarithm.

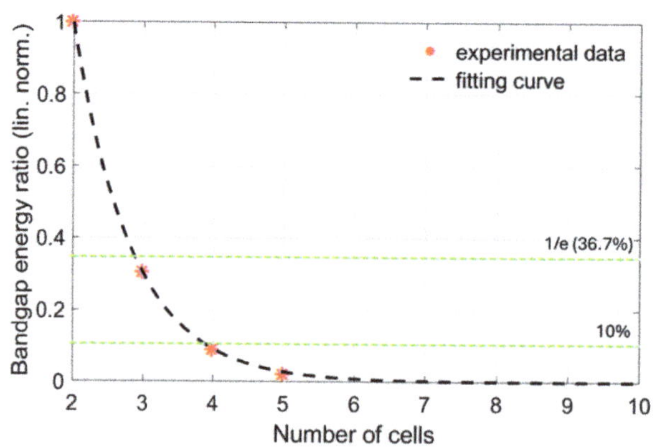

Figure 4. Decay of the energy of the first band gap with increasing cell multiplicity. The horizontal green lines mark the 1/e (36.7%) and 10% energy drop criteria, with respect to the 2-cell structure.

The resulting 4-points graph is fitted by an exponential decay function, expressible as the following:

$$g(n) = a\exp(-bn) \tag{8}$$

with $a = 11,\ b = 1.2$, which was found to perfectly match the observed bandgap energy decay. The results are presented in Figure 4 (see below), showing that 3-cell structures already fulfill the $1/e$ energy drop criterion. Even for a stricter criterion of 10% (-10 dB), 4 cells are sufficient.

Moreover, these experimental results reveal the impact of cell multiplicity on the passband frequency regions. Particularly, the first passband (~2.5–3.5 kHz) appears as a mainly uniform and broadened spectral feature. Distinct spectral peaks can be vaguely observed in some of the responses. In the second passband, frequency range (~9.5–10.9 kHz), distinct resonant peaks are clearly observed which are equal in number to the unit cells of the structure, as shown in Figure 5. In the third passband (~13.5–19.5 kHz), the behaviour of the resonant peaks is more complex, where the number of peaks N appears to relate to the unit cell number n according to the formula $N = 2n - 1$. Notably, the experimental findings of Figure 5 show that by increasing the number of unit cells, the spectral flatness of the passband regions increases. Here, as a measure of spectral flatness, the average peak-to-dip distance is taken, which approximately is 10 dB for the 2-cell structure while it reduces to ~3 dB in the 5-cell structure. This is because the contribution of the increased number of resonant peaks leads to spectral overlaps that shape a flat spectrum. It should be mentioned here that the high-pass slope observed in the passband regions, as for example in the 4-cell and 5-cell structures, can be attributed to the high pass profile of the LPSS excitation and is not an inherent acoustic feature of the structure. Hence, the slope would be absent in the structure response to a flat acoustic source. A thorough interpretation of the physical mechanisms leading to the formation of these features is given in the Discussion.

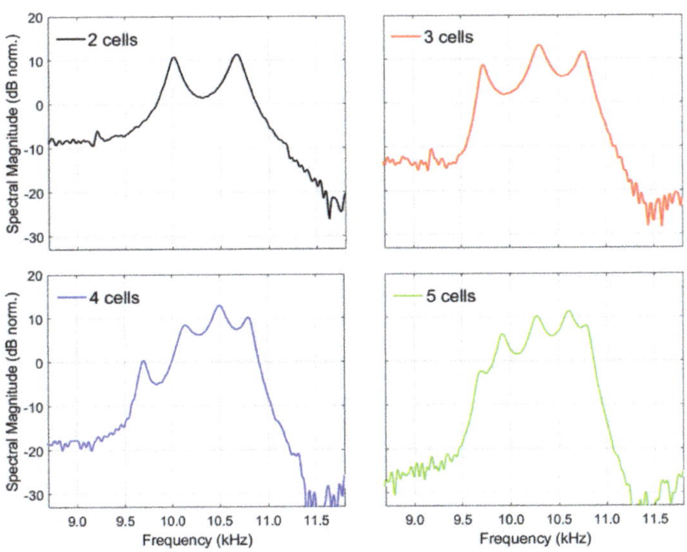

Figure 5. Details of the measured spectral profiles shown in Figure 3a for the frequency range 8.5–12 kHz. The resonant peaks of the second passband are equal in number with the number of the unit cells.

Furthermore Figure 3b demonstrates the excellent agreement between the computational model and the experimental results in terms of the relation of the bandgap depth and the number of resonant peaks of the passband regions. Minor deviations can be attributed to the fact that the computational model simulates the receiver as a point with negligible dimensions in contrast to the real microphone, which effectively is a surface. On the microphone, the more complex superposition of wave fronts exiting the PC leads to deeper bandgaps, especially as the number of unit cells increases.

4. Discussion

Here we present an analysis of the formation of the transmission spectral peaks from the interaction of the propagating acoustic wave and the cavities and conduits forming the internal structure of the PCs. Specifically, the acoustic impedance, or equivalently the acoustic refractive index of the air encountered by the propagating acoustic waves, alternate successively between the narrow cylindrical conduits and the comparatively wider spheres within the PC structure. As a result, the acoustic behavior of the PCs resembles that of a layered material, where sound waves at each interface between successive layers are partially reflected, in a process resembling Bragg scattering in crystals. The reflected waves then interfere with each other constructively (in phase) or destructively (out of phase) forming spectral regions where propagation is enhanced or suppressed respectively. In the passband regions, Helmholtz resonances appear in the spherical cavities while standing waves appear in the cylindrical conduits, especially in the high end of the audible range. The resonances in the spherical cavities can be estimated by the spherical harmonic functions $Y_l^m(\theta,\varphi)$, which are given by the following equation [40]:

$$\nabla^2 Y_l^m(\theta,\varphi) + l(l+1)Y_l^m(\theta,\varphi) = 0 \quad (9)$$

where θ and φ represent the azimuth and elevation angles, respectively, while l and m represent the degree and order of the harmonic function, respectively. A schematic representation of the pressure distribution inside a spherical cavity for the first three degrees (l = 0, 1 and 2) of the spherical harmonic is shown in Figure 6. The different colours in the geometrical distributions represent out of phase pressure profiles.

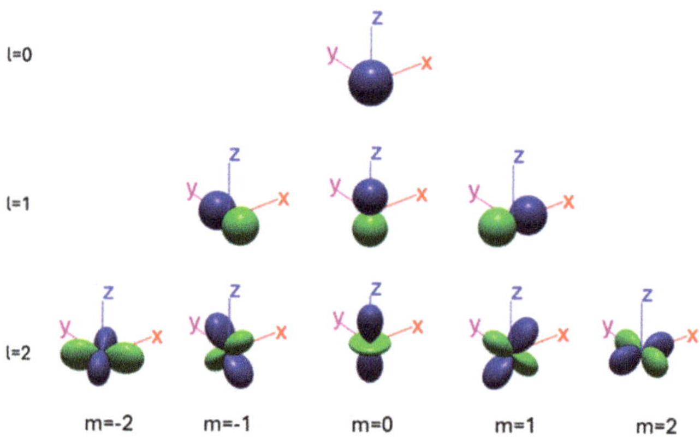

Figure 6. Geometrical shapes of spherical harmonics for l = 0, 1, 2 and m = 0, ±1, ±2, taken from [41]. The different colors indicate opposite phase of the acoustic pressure.

Figure 7a shows the transmission spectrum for the unit cell, where three resonant peaks appear with central frequencies of 3.1 kHz, 10.5 kHz and 16.3 kHz. In accordance with the simulated mode shapes (Figure 7b), namely the simulated pressure distribution at a particular frequency, these spectral peaks can be attributed to the spherical harmonics arising in the spherical cavity. Particularly, the first peak (I) corresponds to the zero-degree spherical harmonic with $l = 0$, while the second (II) and third peaks (III) correspond to the first (l = 1) and second (l = 2) degree spherical harmonics respectively. The symmetry conditions of the studied scenario are dictated by the source, which is placed before the left hole of the structure, leading to an acoustic wave that initially propagates along the x axis (see Figure 7c). This results in a symmetric behaviour of the structure in the y and z directions, as can be observed by the two-plane representations of the figure (Figure 7c). It should be noted that the mode shapes shown in Figure 7b,c represent absolute pressure. Hence, they do not contain any phase information, in contrast to the graph of the spherical harmonics (Figure 6), in which the opposite phase is denoted with different colours.

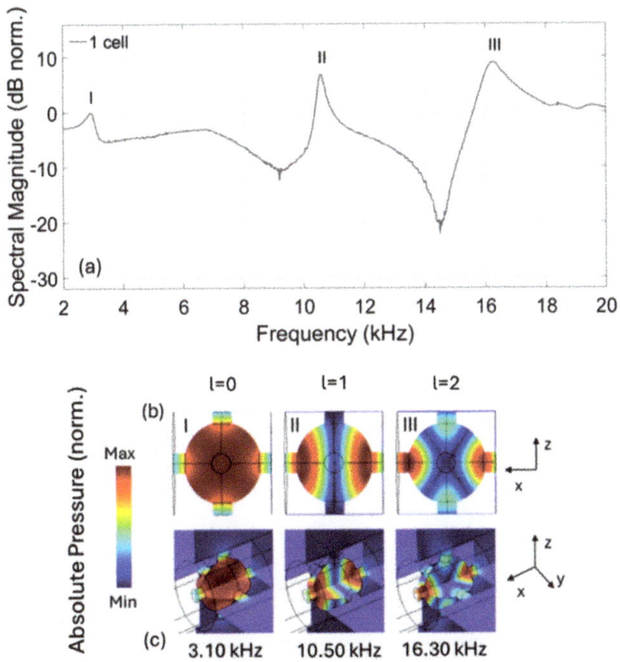

Figure 7. (**a**) Experimentally measured transmission spectrum of the unit cell, with I, II and III indices denoting the $l = 0, l = 1$ and $l = 2$ resonances, respectively. (**b**) 2D and (**c**) pseudo-3D representations of the absolute pressure distribution (mode shapes) for the three resonant frequencies and $l = 0,\ 1,\ 2$ denote the first three degrees of the spherical harmonics.

The fundamental resonance frequency of a spherical cavity with six narrow conduits can be calculated from the following equation [42]:

$$f_0 = \frac{u_{\mathrm{air}}}{2\pi}\sqrt{\frac{6A_{\mathrm{c}}}{V_{\mathrm{c}}L_{\mathrm{c,eff}}}} \tag{10}$$

$$L_{\mathrm{c,eff}} = L_{\mathrm{c}} + 1.7r \tag{11}$$

where $A_{\mathrm{c}} = \pi r^2$ is the conduit cross sectional area, u_{air} is the speed of sound in ambient air, V_{c} is the volume of the spherical cavity, L_{c} is the conduit length, $L_{\mathrm{c,eff}}$ is the effective conduit length and r the conduit diameter. For the geometry of the unit cell under investigation, the fundamental frequency is $f_0 = 2.85$ kHz, which is in accordance with the experimentally measured ($f_0 \sim 3$ kHz) and numerically calculated frequency.

Regarding the 2-cell PC, the transmission spectrum shown in Figure 8a features three passband regions, while Figure 8b corresponds to the simulated mode shapes. Here, the first passband features two resonant peaks (I and II), while the second passband feature two prominent resonant peaks (III and IV) and the third features three (V, VI and VII). The appearance of additional resonances can be attributed to the coupling between the two cavities via a conduit with a length that is double the length of a single conduit. Particularly the resonant peaks of the first passband correspond to the symmetric and antisymmetric modes of the zero-degree spherical harmonic (l = 0) of the spherical cavities, coupled via the cylindrical conduit. The frequencies of the symmetric and antisymmetric modes exhibit a small frequency shift from the fundamental frequency f_0 due to the coupling factor. The same underlying mechanism also forms the two resonant peaks of the second passband, where the spherical harmonic forming the resonant peaks in this case is the first degree ($l = 1$). In the third passband, the resonant peaks IV and V again correspond to the symmetric and antisymmetric modes of the second degree harmonic ($l = 2$). The third resonant peak with experimentally measured frequency, 18.35 kHz, however, can be attributed to the conduit itself acting as a resonator. Particularly, this conduit resonance corresponds of the $\lambda/2$ standing wave condition, taking into account the length and the end correction for the conduit geometry. Calculation of the conduit resonant frequency using the Equation (7) gives $f_c = 18.3\text{kHz}$, expressible as the following:

$$f_c = 2\frac{u_{air}}{L_{c,\text{eff}}} \tag{12}$$

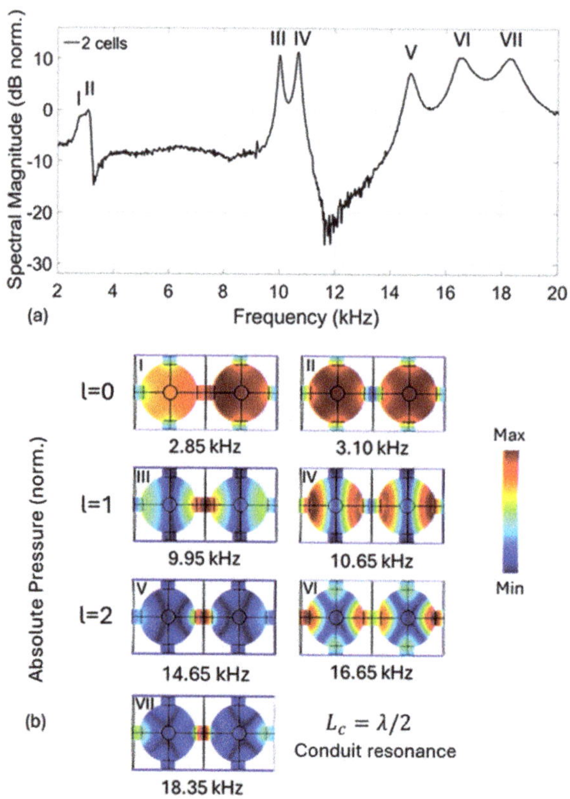

Figure 8. (**a**) Experimentally, measured transmission spectrum of the 2-cell structure, with I, II and III indices denoting the resonance peaks of the first passband. The III and IV indices denote the resonance peaks of the second passband and the V, VI and VII indices denote the resonance peaks of the third passband. (**b**) 2D representation of the absolute pressure distribution (mode shapes), l = 0, 1 and 2 denote the first three degrees of the spherical harmonics, and λ is the longest wavelength for which the standing wave condition in the conduit is fulfilled.

For a larger number of unit cells, the number of resonant peaks within the first and second passbands are equal to the unit cell multiplicity, with each peak associated with symmetric and antisymmetric modes, respectively. However, due to the very small Δf for the harmonic of the first degree, the peaks unify, forming a single and widened peak. Finally, for the third passband, the number of resonant peaks follows the relation $N = 2n - 1$. In this relation, n resonances pertain to the spheres acting as resonators, with their symmetric and antisymmetric modes, which are equal to the number of the spheres. The additional resonances originate from the $n - 1$ conduits acting as resonators. The conduits also constitute a system of interconnected resonators, having symmetric and antisymmetric modes. Since the number of modes equal the number of conduits, they are $n - 1$. By adding the sphere and conduit modes, we get the total $2n - 1$ resonant peaks observed in the 3rd passband both in the measurements and simulations.

5. Conclusions

In this work we studied the impact of the multiplicity of unit cells on the profile of the acoustic response of PCs. We presented our experimental results and simulations demonstrating that by varying the number of unit cells it is possible to manipulate the PC spectral profiles. It was shown that an increase in the multiplicity of unit cells results in an increase in the depth of the first and second bandgap regions. Specifically, the acoustic energy in the first band gap decays exponentially with the number of cells. Based on this finding, a methodology to estimate the number of unit cells required for the structure to approximate the behavior of an infinite structure was outlined. Furthermore, it was shown that cell multiplicity also determines the number of resonant peaks in the passband regions. With the help of computationally evaluated mode shapes, the resonant peaks were attributed to distinct spherical harmonics of the spherical cavities and standing wave modes of the cylindrical conduits. This relation was found to be in accordance with calculations of the resonant frequencies using the analytical formulas for the spherical harmonics and standing waves. Notably, it was shown that an increase in the unit cell multiplicity leads to an increased spectral flatness of the passband region. These results provide a clear demonstration of the controllability of the acoustic response of PCs by modulating their unit cell multiplicity. They allow for the determination of the number of unit cells in order to achieve the desired (a) acoustic energy suppression in the bandgaps and (b) spectral flatness in the passband regions. Moreover, they contribute to the understanding of the functionality of such PC structures as well as to the systematization of their design process. The control of the acoustic wave transmission via PCs paves the way for innovative applications in the field of acoustics, such as noise insulation and room acoustics.

In the future, the presented methodology will be applied to other PC structures, including structures with defected cells and higher-order structures. The results will be compared to the findings presented here, to identify potential global patterns in the behavior of PCs. Also, an investigation will be carried out to identify the impact of the material density of the structure on the observed transmission spectra. This will be achieved by measuring identical structures that are 3D printed with varying infill densities of the PLA material via the LPSS method.

Author Contributions: Concept: K.K.; Acoustic measurements: E.K.K. and K.K.; Data analysis: E.K.K., K.K. and V.D.; Analysis software: K.K.; Simulations: N.A.-Z.; Interpretation: K.K., E.K.K. and N.A.P.; Sample design: M.S. and N.A.-Z.; Sample development: D.T.G.K. and N.A.-Z. First draft: E.K.K. and K.K.; Resources: M.T., N.A.P. and M.B.; Project Administration: M.T., N.A.P. and M.B.; Supervision: N.A.P.; All authors revised and edited the manuscript. All authors have read and agreed to the published version of the manuscript.

Funding: This research received no external funding.

Institutional Review Board Statement: Not applicable.

Informed Consent Statement: Not applicable.

Data Availability Statement: The original contributions presented in the study are included in the article. Further inquiries can be directed to the corresponding author.

Conflicts of Interest: The authors declare no conflicts of interest.

References

1. Christensen, J.; Javier Garcia de Abajo, F. Negative refraction and backward waves in layered acoustic metamaterials. *Phys. Rev. B* **2012**, *86*, 024301. [CrossRef]
2. Chen, H.; Chan, C.T. Acoustic cloaking in three dimensions using acoustic metamaterials. *Appl. Phys. Lett.* **2007**, *91*, 183518. [CrossRef]
3. Clémentine, L. *Finite Element Modelling of Acoustic Shielding via Phononic Crystal Structures*; Semester Project; Ecole Politechnique Federale de Lausanne: Lozan, Switzerland, 2016.
4. Kee, C.S.; Kim, J.E.; Park, Y.H.; Chang, J.K.; Lim, H. Essential role of impedance in the formation of acoustic bandgaps. *J. Appl. Phys.* **2000**, *87*, 1593. [CrossRef]
5. Lu, M.H.; Feng, L.; Chen, Y.F. Phononic crystals and acoustic metamaterials. *Mater. Today* **2009**, *12*, 34–42. [CrossRef]
6. Fomenko, S.I.; Golub, M.V.; Alexandrov, A.A.; Chen, A.L.; Wang, Y.S.; Zhang, C. Band-gaps and low transmission pass-bands in layered piezoelectric phononic crystals. In Proceedings of the International Conference DAYS on DIFFRACTION, St. Petersburg, Russia, 27 June–1 July 2016.
7. Sabiju, V.V.; Alejandro, M.A.; Hans, G. Phononic crystals band gap manipulation via displacement modes. *Solid State Commun.* **2023**, *361*, 115061.
8. Croënne, C.; Lee, S.J.E.; Hu, H.; Page, H.J. Band gaps in phononic crystals: Generation mechanisms and interaction effects. *AIP Adv.* **2011**, *1*, 041401. [CrossRef]
9. Aravantinos, Z.N.; Sigalas, M.M.; Katerelos, G.T.D. Complete acoustic bandgaps in a three-dimensional phononic metamaterial with simple cubic arrangement. *J. Appl. Phys.* **2023**, *133*, 065101. [CrossRef]
10. Khelif, A.; Adibi, A. *Phononic Crystals Fundamentals and Applications*; Springer: New York, NY, USA, 2016.
11. Deymier, A.P. *Acoustic Metamaterials and Phononic Crystals*; Springer: Berlin/Heidelberg, Germany, 2013.
12. Han, D.; Zhang, G.; Zhao, J.; Yao, H.; Liu, H. Study on Band Gap and Sound Insulation Characteristics of an Adjustable Helmholtz Resonator. *Appl. Sci.* **2022**, *12*, 4512. [CrossRef]
13. Zhang, X.; Qu, Z.; Wang, H. Enhanced sound insulation performance by a staggered phononic crystal with fins. *Appl. Acoust.* **2023**, *208*, 109398. [CrossRef]
14. Zhang, X.C.; Qu, Z.G.; He, X.C.; Lu, L.D. Experimental study on the sound absorption characteristics of continuously graded phononic crystals. *AIP Adv.* **2016**, *6*, 105205. [CrossRef]
15. Yang, S.; Chang, H.; Wang, Y.; Yang, M.; Sun, T. A phononic crystal suspension for vibration isolation of acoustic loads in underwater gliders. *Appl. Acoust.* **2024**, *216*, 109731. [CrossRef]
16. Pan, Y.; Liu, R.; Bin, G.; He, X. Vibration and noise reduction of phononic crystal structure laid on the transmission path of axial piston pump. *Appl. Acoust.* **2022**, *200*, 109075. [CrossRef]
17. Aravantinos, Z.N.; Sigalas, M.M.; Katerelos, G.T.D. Phononic metamaterial for efficient sound attenuation applications. *Build Acoust.* **2022**, *29*, 203–220. [CrossRef]
18. Liu, R.; Shu, H.; Wang, X. Research on Soundproof Properties of Cylindrical Shells of Generalized Phononic Crystals. In Proceedings of the 5th International Conference on Computer-Aided Design, Manufacturing, Modelling and Simulation, Busan, Republic of Korea, 22–23 April 2017.
19. Laude, V. Principles and properties of phononic crystal waveguides. *APL Mater.* **2021**, *9*, 080701. [CrossRef]
20. Khelif, A.; Djafari-Rouhani, B.; Vasseur, O.J.; Deymier, A.P. Transmission and dispersion relations of perfect and defect-containing waveguides structures in phononic band gap materials. *Phys. Rev. B* **2003**, *68*, 024302. [CrossRef]
21. Olson, R.H.; El-Kady, I.; Mehmet, F.S.; Melanie, R.T.; McCormick, F.B.; Fleming, G.J. *Research on Micro-Sized Acoustic Bandgaps Structures*; SAND2010-0044; Sandia National Laboratories (SNL): Albuquerque, NM, USA; Livermore, CA, USA, 2010.
22. Nomura, M.; Laude, V.; Maldovan, M. Phononic crystals at various frequencies. *APL Mater.* **2022**, *10*, 050401. [CrossRef]
23. Mukhin, N.; Kutia, M.; Aman, A.; Steinmann, U.; Lucklum, R. Two-Dimensional Phononic Crystal Based Sensor for Characterization of Mixtures and Heterogeneous Liquids. *J. Sens.* **2022**, *22*, 2816. [CrossRef]
24. Aravantinos-Zafiris, N.; Sigalas, M.M. Bandgaps in Phononic strip waveguides. *J. Appl. Phys.* **2012**, *111*, 123516. [CrossRef]
25. Chen, Z.G.; Zhao, J.; Mei, J.; Wu, Y. Acoustic frequency filter based on anisotropic topological phononic crystals. *Sci. Rep.* **2017**, *7*, 15005. [CrossRef]
26. Pennek, Y.; Bahram, D.R.; Hocine, L.; Jerome, V.; Hladky-Hennion, A.C. Phononic crystals and manipulation of sound. *Phys. Status Solidi C* **2009**, *6*, 2080–2085. [CrossRef]

27. Konstantopoulou, A.; Aravantinos-Zafiris, N.; Sigalas, M.M. Tunable bandgaps and defect states in 3D phononic hollow structures. *J. Appl. Phys.* **2020**, *117*, 075101. [CrossRef]
28. Ma, F.; Huang, Z.; Liu, C.; Wu, J.H. Acoustic focusing and imaging via phononic crystal and acoustic metamaterials. *J. Appl. Phys.* **2022**, *131*, 011103. [CrossRef]
29. Ji, H.W.; Yang, F.; Qi, A.Q.; Wu, X.; Lu, B.; Ni, J. Tunable phononic crystal waveguides based on the double tuning method. *AIP Adv.* **2022**, *12*, 025018. [CrossRef]
30. Khelif, A.; Choujaa, A.; Benchabane, S.; Djafari-Rouhani, B.; Laude, V. Guiding and bending of acoustic waves in highly confined phononic crystal waveguides. *Appl. Phys. Lett.* **2004**, *84*, 4401. [CrossRef]
31. Maldovan, M. Sound and heat revolutions in phononics. *Nature* **2013**, *503*, 209–217. [CrossRef]
32. Li, Y.; Yu, G.; Liang, B.; Zou, X.; Li, G.; Cheng, S.; Cheng, J. Three dimensional Ultrathin Planar Lenses by Acoustic Metamaterials. *Sci. Rep.* **2014**, *4*, 6830. [CrossRef]
33. Kaleris, K.; Kaniolakis-Kaloudis, E.; Aravantinos-Zafiris, N.; Katerelos, T.G.D.; Dimitriou, M.V.; Bakarezos, M.; Tatarakis, M.; Mourjopoulos, J.; Sigalas, M.M.; Papadogiannis, A.N. Acoustic metamaterials characterization via laser plasma sound sources. *Commun. Mater.* **2024**, *5*, 93. [CrossRef]
34. Kaleris, K.; Orfanos, Y.; Bakarezos, M.; Papadogiannis, N.; Mourjopoulos, J. Experimental and analytical evaluation of the acoustic radiation of femtosecond laser plasma filament sound sources in air. *J. Acoust. Soc. Am.* **2019**, *146*, 212–218. [CrossRef]
35. Kaleris, K.; Orfanos, Y.; Petrakis, S.; Bakarezos, M.; Tatarakis, M.; Mourjopoulos, J.; Papadogiannis, N.A. Laser-plasma sound sources in atmospheric air: A systematic experimental study. *J. Sound Vib.* **2023**, *570*, 118000. [CrossRef]
36. Oksanen, M.; Hietanen, J. Photoacoustic breakdown sound source in air. *Ultrasonics* **1994**, *32*, 327–331. [CrossRef]
37. Gómez Bolaños, J.; Pulkki, V.; Karppinen, P.; Hæggström, E. An optoacoustic point source for acoustic scale model measurements. *J. Acoust. Soc. Am.* **2013**, *133*, 221–227. [CrossRef] [PubMed]
38. Kaleris, K.; Tazes, I.; Orfanos, Y.; Petrakis, S.; Bakarezos, M.; Mourjopoulos, J.; Dimitriou, V.; Tatarakis, M.; Papadogiannis, N.A. Experimentally validated modeling of the optical energy deposition in highly ionized ambient air by strong femtosecond laser pulses. *Eur. Phys. J. D* **2021**, *75*, 236. [CrossRef]
39. Audacity. 2025. Available online: https://www.audacityteam.org/ (accessed on 8 January 2025).
40. Poletti, M. Unified description of ambisonics using real and spherical harmonics. In Proceedings of the Ambisonic Symposium, Graz, Austria, 24–27 June 2009.
41. Archontis Politis. Real/Complex Spherical Harmonic Transform, Gaunt Coefficients and Rotations. GitHub. Available online: https://github.com/polarch/Spherical-Harmonic-Transform (accessed on 4 February 2025).
42. Langfeldt, F.; Hoppen, H.; Gleine, W. Resonance frequencies and sound absorption of Helmholtz resonators with multiple necks. *Appl. Acoust.* **2019**, *145*, 314–319. [CrossRef]

materials

MDPI

Article

Prediction and Optimization of the Long-Term Fatigue Life of a Composite Hydrogen Storage Vessel Under Random Vibration

Xiaoshuang Xiong [1,2], Wentao Wang [2], Xiang Li [3,*], Fei Fan [1], Jiacheng Zhou [1,2,*] and Mingzhang Chen [4]

1 Hubei Key Laboratory of Digital Textile Equipment, Wuhan Textile University, Wuhan 430200, China; xsxiong@wtu.edu.cn (X.X.); ffan@wtu.edu.cn (F.F.)

2 School of Mechanical Engineering & Automation, Wuhan Textile University, Wuhan 430200, China; 2215373006@wtu.edu.cn

3 China Special Equipment Inspection and Research Institute, Beijing 100029, China

4 Hubei Key Laboratory of Advanced Technology for Automotive Components, Wuhan University of Technology, Wuhan 430070, China; chenmingzhang@whut.edu.cn

* Correspondence: lixiang@csei.org.cn (X.L.); 2022021@wtu.edu.cn (J.Z.)

Abstract: A composite hydrogen storage vessel (CHSV) is one key component of the hydrogen fuel cell vehicle, which always suffers random vibration during transportation, resulting in fatigue failure and a reduction in service life. In this paper, firstly, the free and constrained modes of CHSV are experimentally studied and numerically simulated. Subsequently, the random vibration simulation of CHSV is carried out to predict the stress distribution, while Steinberg's method and Dirlik's method are used to predict the fatigue life of CHSV based on the results of stress distribution. In the end, the optimization of ply parameters of the composite winding layer was conducted to improve the stress distribution and fatigue life of CHSV. The results show that the vibration pattern and frequency of the free and constrained modes of CHSV obtained from the experiment tests and the numerical predictions show a good agreement. The maximum difference in the value of the vibration frequency of the free and constrained modes of CHSV from the FEA and experiment tests are, respectively, 8.9% and 8.0%, verifying the accuracy of the finite element model of CHSV. There is no obvious difference between the fatigue life of the winding layer and the inner liner calculated by Steinberg's method and Dirlik's method, indicating the accuracy of FEA of fatigue life in the software Fe-safe. Without the optimization, the maximum stresses of the winding layer and the inner liner are found to be near the head section by 469.4 MPa and 173.0 MPa, respectively, and the numbers of life cycles of the winding layer and the inner liner obtained based on the Dirlik's method are around 1.66×10^6 and 3.06×10^6, respectively. Through the optimization of ply parameters of the composite winding layer, the maximum stresses of the winding layer and the inner liner are reduced by 66% and 85%, respectively, while the numbers of life cycles of the winding layer and the inner liner both are increased to 1×10^7 (high cycle fatigue life standard). The results of the study provide theoretical guidance for the design and optimization of CHSV under random vibration.

Keywords: composite hydrogen storage vessel (CHSV); random vibration; free and constrained modes; fatigue life; numerical simulation

1. Introduction

Hydrogen energy is extensively considered the most promising candidate to replace fossil fuels in the field of transportation, with the advantages of zero greenhouse gas emissions, no pollution, and renewability, and it is especially widely used in long-distance and heavy-duty trucks [1]. CHSV is one key component of the hydrogen fuel cell vehicle, which is directly related to the driving range, production cost, and safety of the vehicle [2]. During the actual vehicle driving cycle, a lot of uncertainty, such as braking and turbulence, occur randomly, which will cause stress concentration and excessive deformation at the connection of each component of the CHSV, resulting in vibration fatigue failure and a reduction in service life [3]. Therefore, it is of great importance to carry out the random vibration damage analysis, fatigue life prediction, and structure optimization of CHSV for the long-term operation of hydrogen fuel cell vehicles.

In the past few decades, much research has been conducted on the structural design and optimization of CHSV to improve safety performance. Bouhala et al. summarized relevant simulation models and proposed innovative methods for designing CHSV with high performance [4]. A finite element model was developed by Moskvichev et al. [5] to simulate the stress-strain behavior of CHSV with different numbers of composite layers under internal pressure. Their work investigated the effect of the propagation of initial defects in composite shells and the critical size of surface cracks. Kabiri et al. [6] implemented the Tsai-Hill failure criterion into the finite element model of CHSV to study the stress distribution and the critical region and then investigated its fatigue life using the residual strain energy method. Vafaeesefat et al. [7] proposed a multi-objective strategy for the head shape and winding angle optimization of CHSV, taking into account the internal pressure and volume, the vessel weight, and the composite material properties. Furthermore, Wang et al. [8] further proposed a prediction method for the CFRP CHSV under cyclic fatigue loading and high-temperature conditions based on the integration of micromechanics of failure and the time-temperature superposition principle method. Roh et al. [9] established a finite element model of CHSV to analyze the stress distribution among the vessels with a new integrated end-cap design for the dome section. They found that the high-stress points shifted from the dome to the cylindrical section of the vessel with the integrated end-cap design. In addition, Cho et al. [10] further investigated the effect of dome curvature on the failure mode of CHSV through hydraulic burst tests and numerical simulations. To study the complex phenomenon of plastic liner blistering during the depressurization of type IV hydrogen storage tanks, Yersak et al. [11] developed a polymer liner blistering model, which greatly reduced the experimental cost. Halm et al. [12] developed a finite element model to simulate the coupled effects of mechanical damage and temperature on the burst and leak of type IV hydrogen storage tanks under fire tests, which can accurately predict the time to burst of the composite tank, as well as the transition between burst and leak. Rohit et al. [13] conducted a comprehensive investigation into the structural and explicit performance of three types of CHSVs using 3D models for drop and crash tests. The study revealed that Type 1 CHSVs failed to meet safety requirements under extreme conditions, while Type 3 CHSVs exhibited induced stresses below the yield point in both scenarios, and Type 4 CHSVs presented the best performance. Li et al. [14] investigated the hydrogen fill rate, ambient temperature, volume, and hydrogen inlet temperature to evaluate their effects on the temperature rise within the CHSV. Zhou et al. [15] conducted impact damage and bursting experiments on Type III and Type IV cylinders and found that with the increase in impact energy, the fiber fracture area, matrix crack area, and impact depth of the two cylinder types increased. Furthermore, in the same impact energy, Type III cylinders presented a smaller impact depth but a larger fiber fracture and matrix crack area. Hu et al. [16] conducted finite element modeling of CHSV based on the Hashin damage model and metal plastic damage theory [17,18] and carried out blasting tests. Their studies showed that the progressive damage model can effectively investigate the matrix cracking and fiber fracture and predict burst pressure.

Although the finite element modeling and experiment tests have been applied successfully to the structural design and optimization of CHSV, most of them are focused on the performance analysis and service life prediction of CHSV under the inter pressure condition, the leaking condition, and the fire condition. Nevertheless, the CHSV often inevitably suffers vibrations from uneven road surfaces and internal vibrations during vehicle operation. Therefore, it is imperative and of great importance to study the dynamic response and fatigue life analysis of CHSV under random vibration. Yao et al. [19] proposed an approach to optimize the strapping parameters for in-vehicle CHSV using a GA-XGBoost model, and finite element models were established to verify the optimized results. Huang et al. [20] used the finite element method to conduct the multi-objective optimization of stiffness and strength of a two bottle horizontal hydrogen supply system under vibration conditions. Coskun et al. [21] conducted a study on CHSV with various geodesic dome profiles, analyzing modal and random vibration responses. They examined five polar opening radii to reveal the effects of filament winding orientation and internal pressure severity on the dynamic characteristics. Ryu et al. [22] analyzed the effect of the hydrogen shaking phenomenon inside the CHSV by performing the frequency characteristics of the CHSV at zero frequency, intrinsic frequency, and train operation. As the frequency increased, the free surface wave of liquid hydrogen became shorter and more pronounced. Wang et al. [23] further investigated the intrinsic frequency and mode shapes of CHSV, adopted four calculation methods to estimate the vibration fatigue life of the CHSV inner liner, and found that the number of cycle times of the CHSV inner liner calculated by the four methods all reached over 10^6 cycles. It is not difficult to see from the above research that the structural optimization of the hydrogen supply system and the fatigue analysis of the inner liner of CHSV under random vibration have been widely discussed, but few studies have been conducted on the fatigue analysis and structural optimization of the composite winding layer of CHSV under random vibration.

In this paper, finite element models were developed to simulate the dynamic response of CHSV under random vibration. Modal testing experiments were conducted to verify the accuracy of numerical simulations. Then, the stress and deformation simulation for CHSV under random vibration was performed for the fatigue life prediction, including the stress and deformation analysis of the aluminum liner and composite winding layer. The Steinberg's method and the Dirlik's method were used to predict the fatigue life of CHSV under random vibration. In the end, the optimization of ply parameters of the composite winding layer was conducted to improve the stress distribution and fatigue life of CHSV. The results of the study provide a simulation strategy and an experimental method for the prediction and optimization of fatigue life of CHSV under random vibration.

2. Methods

2.1. Basic Theory of Stochastic Processes with Power Spectral Density (PSD)

Stochastic processes can be classified into two categories according to their statistical properties over time: smooth random vibrations and non-smooth random vibrations. The power spectral density (PSD) can be used to both describe the smooth process in the frequency domain and express the energy distribution of the stochastic process, which is an important parameter of the stochastic process in the frequency domain [24] and is defined as follows:

$$S_x(\omega) = \int_{-\infty}^{+\infty} R_x(\tau) e^{-j\omega\tau} d\tau \tag{1}$$

$$R_x(\tau) = \frac{1}{2\pi} \int_{-\infty}^{+\infty} S_x(\omega) e^{j\omega\tau} d\omega \tag{2}$$

where $S_x(\omega)$ is the self-power spectral density function, ω and f are the circular frequency and frequency, respectively, and $R_x(\tau)$ is the autocorrelation function.

In the engineering application, the circular frequency $\omega > 0$; therefore, the unilateral power spectral density is defined as follows:

$$G(\omega) = \begin{cases} 2S_x(\omega), & \omega > 0 \\ S_x(0), & \omega = 0 \end{cases} \tag{3}$$

The PSD and the root mean square (RMS) of von Mises stress in the software Abaqus 2017 are calculated based on the reference [25]. In the software Abaqus 2017, the PSD for von Mises stress at node a is given as follows:

$$S^a_{mises}(f) = \sum_{\beta=1}^{m} \sum_{\alpha=1}^{m} S_{\alpha\beta}(f) T^a_{\alpha\beta} \tag{4}$$

where m is the number of modes and $S_{\alpha\beta}(f)$ are the elements of the PSD matrix of generalized displacements.

$$T^a_{\alpha\beta} = [\psi^a_\alpha]^T [A] \left[\psi^a_\beta\right] \tag{5}$$

$[\psi^a_\alpha]$ are the modal stress components of the αth mode of node a, and the constant matrix A is given as follows:

$$\begin{pmatrix} 1 & -1/2 & -1/2 & 0 & 0 & 0 \\ -1/2 & 1 & -1/2 & 0 & 0 & 0 \\ -1/2 & -1/2 & 1 & 0 & 0 & 0 \\ 0 & 0 & 0 & 3 & 0 & 0 \\ 0 & 0 & 0 & 0 & 3 & 0 \\ 0 & 0 & 0 & 0 & 0 & 3 \end{pmatrix} \tag{6}$$

Similarly, the root mean square of the von Mises (RMS) stress of node a is computed as

$$RMISES^a(f) = \sqrt{\sum_{\beta=1}^{m} \sum_{\alpha=1}^{m} V_{\alpha\beta}(f) T^a_{\alpha\beta}} \tag{7}$$

where $V_{\alpha\beta}(f)$ are the elements of the variance matrix of generalized displacements.

2.2. Estimation of Fatigue Life Under Random Vibration

Under random vibration conditions, the alternating stress or deformation will cause the fatigue cumulative damage of CHSV [26]. Linear fatigue cumulative damage assumes that the sequence of alternating load and the cumulative damage do not affect each other, which is widely used to predict the fatigue life of structures made of CFRP [27] and aluminum alloy [28]. In addition, the different types of load damage superposition are linear distributions, while fatigue damage occurs when the value of cumulative damage reaches 1. Miner's linear cumulative damage theory is given as follows [29]:

$$D = \sum D_i = \sum \frac{n_i}{N_i} \tag{8}$$

where n_i and N_i denote the actual number of stress cycles and the fatigue life at the stress level S_i, respectively.

In the case that the cyclic stress is continuously varying, Equation (8) can be transformed into

$$D = \int_0^{+\infty} \frac{n_s}{N_s} ds \tag{9}$$

where n_s and N_s are the number of actual cycles and damage cycles when the peak stress is s, respectively. Generally when $D = 1$, the CHSV will be damaged.

The power function model is widely used to describe the fatigue life curve of common composite materials and is given as follows:

$$s^m N = c \tag{10}$$

c and m are the material constants, obtained through fatigue testing. N is the number of damage cycles when the stress amplitude is s.

3. Simulation and Experimental Procedure

3.1. Composite Hydrogen Storage Vessel

Figure 1 shows the CHSV with an inner radius of 74.5 mm and an outer radius of 78.5 mm, which is supported by a Chinese company (Shanghai, China). The CHSV consists of a thin inner liner (h_{Al} = 4 mm) made of aluminum alloy and an outer winding layer (h_{CFRP} = 3.2 mm) made of carbon fiber-reinforced composite (CFRP), which is manufactured through the dry winding process [30]. The stacking sequence of the CFRP winding layer is $[13.5°_2/90°_2]_7$, which consists of long and continuous T700 carbon fibers and FXR-521L/FXC-521L epoxy resin. The viscosity of epoxy resin (supplied by the Shanghai Fuchen Chemical Co., Ltd., Shanghai, China), used for CHSV, is 4000 to 8000 cP at 25 °C, and the solid particle size is less than 10~25 μm. The fiber volume fraction of the CFRP winding layer is approximately 0.62, and the fiber volume fraction of the winding layer is always around 60% to 70% [31]. Material properties of the CHSV are listed in Table 1, which are provided by the Chinese company.

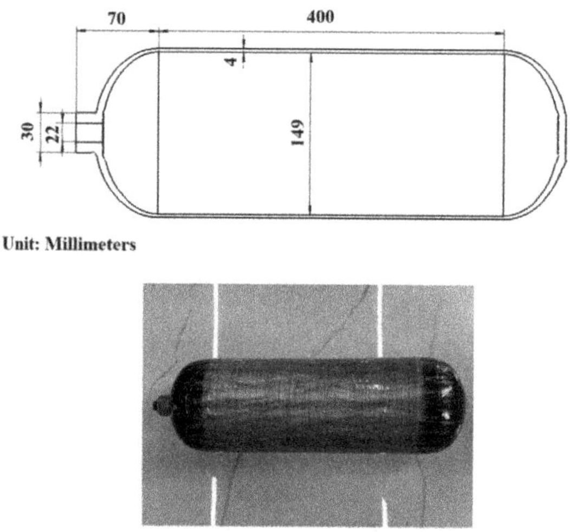

Figure 1. Geometric structure diagram of CHSV.

Table 1. Material properties of the CHSV.

Material Properties	CFRP (T700/Epoxy)	Aluminum Alloy (6061-T6)
Longitudinal modulus, E_1 (GPa)	139.00	6.00
Transverse modulus, $E_2 = E_3$ (GPa)	8.00	
Longitudinal shear modulus, $G_{12} = G_{13}$ (GPa)	6.37	
Transverse shear modulus, G_{23} (GPa)	3.70	
Major Poisson's ratio, $\upsilon_{f12} = \upsilon_{f13}$	0.30	0.30
Minor Poisson's ratio, υ_{f23}	0.35	

3.2. Modal Testing

Modal testing is widely used to investigate the vibrational frequency and mode of a mechanical structure, which is of great significance and the crucial step in studying the dynamic properties of CHSV under random vibration.

In our study, both free modal testing and constrained modal testing were conducted to investigate the vibration modes and frequencies of three same-type CHSVs and verify the accuracy of numerical simulation. As shown in Figure 2A, the CHSV was suspended using four elastic ropes during the free modal test. The equipment for the constrained modal test of CHSV is presented in Figure 2B. The two ends of CHSV were fixed to the brackets by the flexible tapes, and the brackets were attached to the vibration test bench through a bolted connection. Both the free and constrained modal tests were performed using the impact test method, while four PCB three-phase accelerometers were used to collect the acceleration signals of four points evenly distributed around the circumference of CHSV at every hammer blow. In order to accurately obtain the vibration characteristics of CHSV, the acceleration signals of 20 points on five equidistant sections along the CHSV axis were obtained through five impact tests sequentially. The Leuven measurement and system (LMS) data acquisition system was used to obtain and analyze the vibration patterns of CHSV at different intrinsic frequencies.

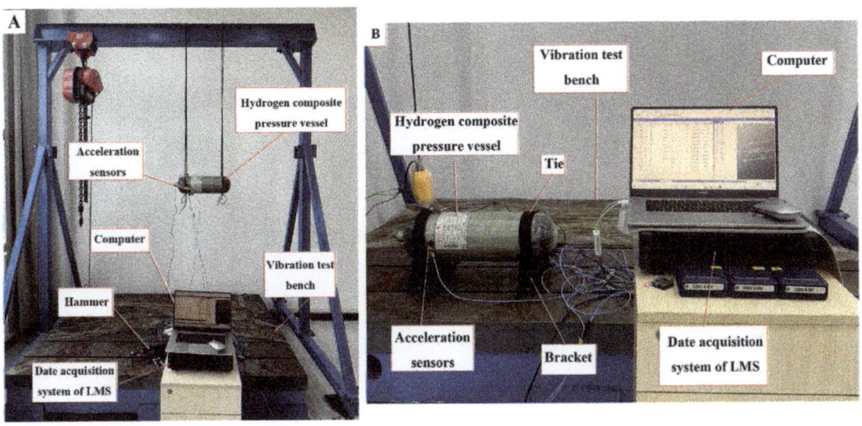

Figure 2. Schematic diagram of (**A**) the free modal experiment and (**B**) the constrained modal experiment of CHSV.

3.3. Finite Element Modeling

Finite element modeling is extensively used in modal analysis and random vibration analysis [32], which can simulate various test schemes and save design time. In the present study, the basically finite element model of CHSV meshed with linear hexahedral elements (C3D8R) was developed in the software Abaqus 2017, which is shown in Figure 3. The linear hexahedral elements (C3D8R) were widely used to simulate the fatigue life of engineering parts under alternating loads, which has the advantages of shortening the computation time, improving the computation accuracy, and preventing the shear self-locking phenomenon [33]. The element number of the inner aluminum alloy liner and CFRP winding layer of CHSV is 72,360 and 720,560, respectively. The geometric dimension of CHSV has been described in Section 3.1, as well as the material properties of the inner aluminum alloy liner and CFRP winding layer, are listed in Table 1.

For the free modal simulation of CHSV, the degrees of freedom for all nodes were set to free, while the degrees of freedom of nodes located at the fixation area of the vessel according to the constrained modal test were constrained for constrained modal simulation. The finite element model and boundary condition of CHSV under the constrained modal test and random vibration condition are presented in Figure 4.

Figure 3. Finite element model of CHSV.

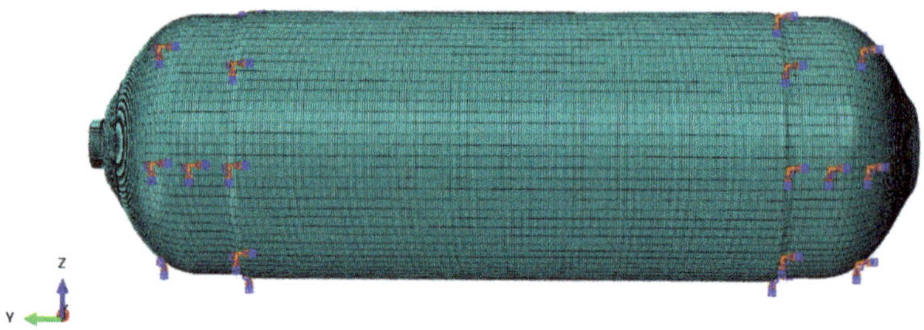

Figure 4. Finite element model and boundary condition of constrained mode and random vibration simulation.

The random vibration simulation of CHSV was then carried out on the basis of constrained modal simulation, and a random vibration load was applied to CHSV. Furthermore, the applied random vibration load generated by a vehicle driving in a real situation was according to the reference [34] and the acceleration power spectral density of the loaded vehicle under a vibration environment is shown in Figure 5. As shown in Figure 5, the vibration in the vertical direction is the most intense, which is more representative and is consistent with the z direction in Figure 4.

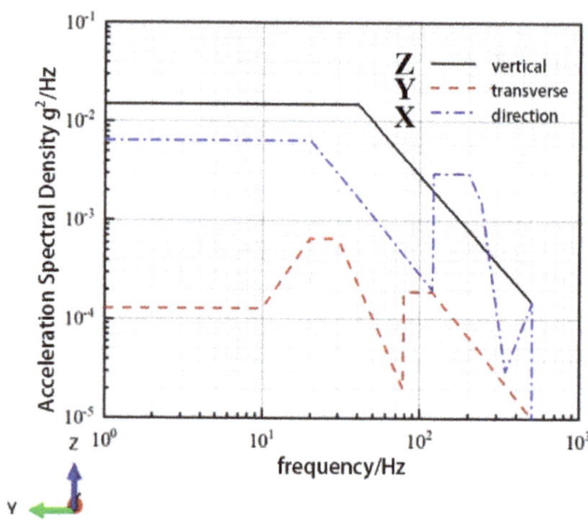

Figure 5. Acceleration power spectral density [34].

4. Results and Discussion

4.1. Modal Simulation and Experiment

Figure 6 compares the finite element predictions and experimental results of the first four orders of free modes of CHSV. There is a good match of vibration pattern and frequency between the results from the finite element analysis (FEA) and from the experiment test. From Figure 6c, it can be observed that the vibration frequency of FEA and the experiment test increases from about 210 Hz to 470 Hz, and the maximum difference in the value of vibration frequency between the results from the FEA and experiment test is about 8.9% at the first-order mode. As shown in Figure 6a,b, the first four vibration patterns of CHSV are, respectively, the compression and bulge that appeared in the middle section with a symmetrical distribution of 90 degrees, the compression and bulge that presented in the middle section with a symmetrical distribution of 45 degrees, the bending in the middle section, and the compression in the middle section.

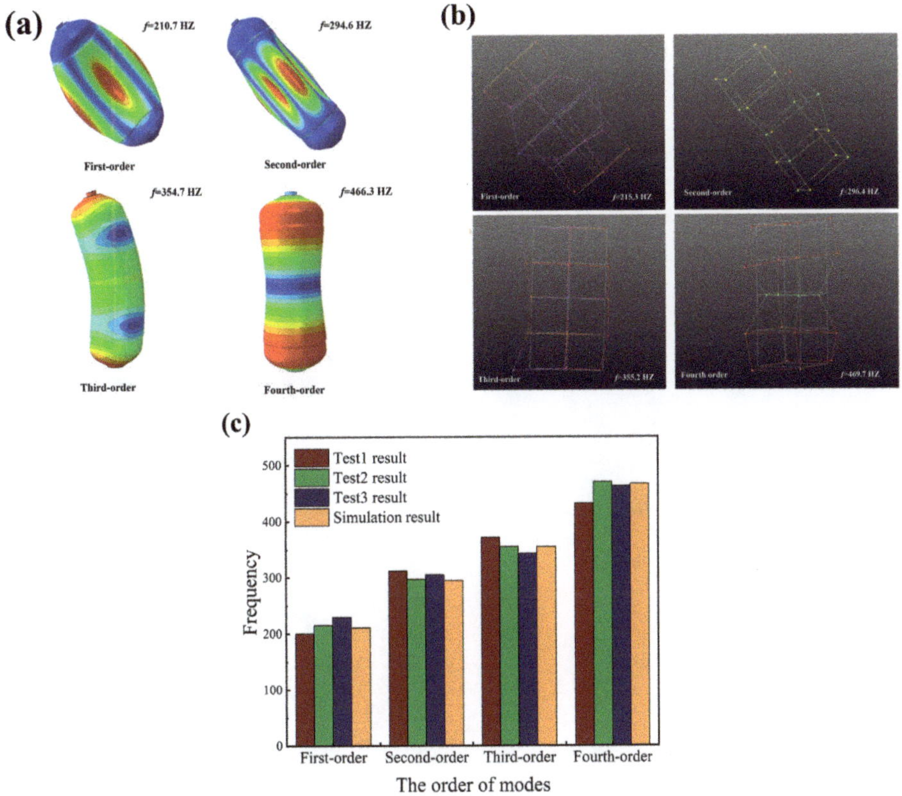

Figure 6. The first four orders of free modes of CHSV: (**a**) the finite element prediction of the vibration modes, (**b**) the experimental result of the vibration modes, and (**c**) the comparison of the vibration frequencies.

Figure 7 shows the comparison between the finite element predictions and experimental results of the first four orders of constrained modes of CHSV. The results of vibration pattern and frequency from FEA show a good agreement with the experimental results. From Figure 7c, it can be found that the vibration frequency of FEA and experiment test increases from about 240 Hz to 580 Hz, and the maximum difference in the value of vibration frequency between the results from the FEA and experiment test is about 8.0% at the first-order mode. Furthermore, the first-order vibration pattern is the cylinder bending in the middle section, the second-order vibration pattern is the expansion of the middle of the cylinder, the third-order vibration pattern is the depression of the upper and lower parts of the cylinder in the opposite direction, and the fourth-order vibration pattern is the twisting of the cylinder in the form of an S shape.

By comparing Figure 6 with Figure 7, it can be found that the same order vibration frequency of the constrained mode is higher than that of the free mode, while their vibration modes of each order show obvious differences. In addition, it can be concluded that the finite element model of CHSV has a high accuracy based on the comparison between the finite element predictions and experimental results of the free mode and constrained mode.

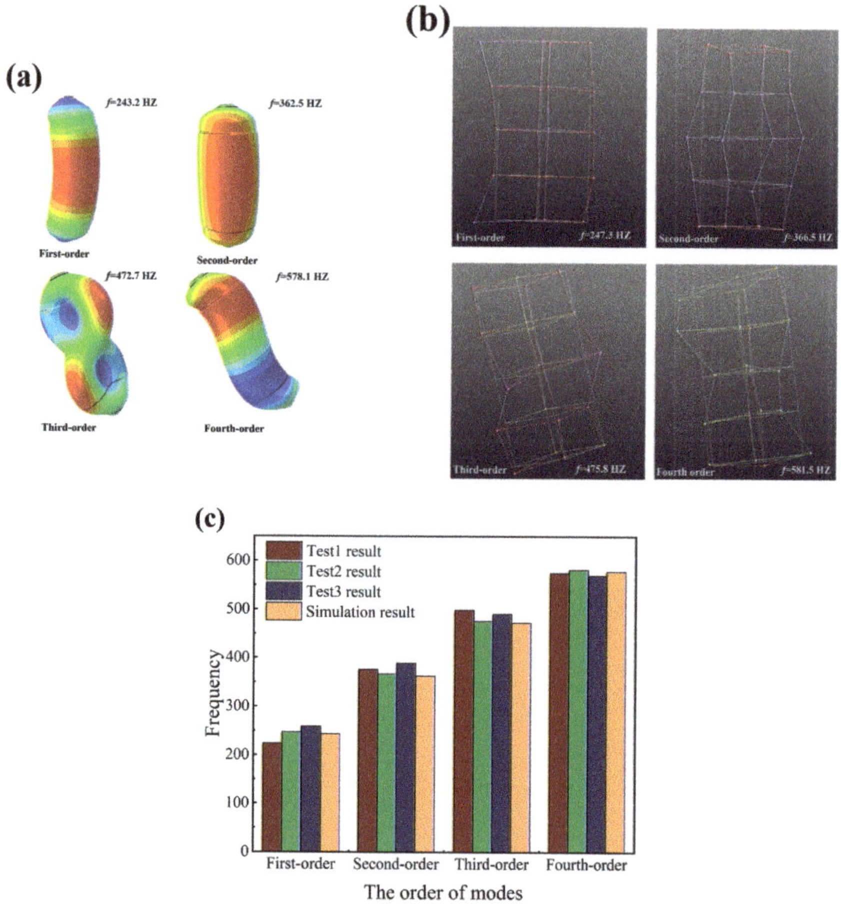

Figure 7. The first four-orders of constrained modes of CHSV: (**a**) the finite element prediction of vibration modes, (**b**) the experimental result of vibration modes, and (**c**) the comparison of vibration frequencies.

4.2. Random Vibration

The von Mises stress of the winding layer and the inner liner of the CHSV under random vibration is shown in Figure 8. It can be seen that the maximum stress points for the winding layer and inner liner are both concentrated near the CHSV mouth, where the location with the thickest fiber buildup is. The maximum value of von Mises stress is 469.4 MPa for the winding layer and 173.0 MPa for the liner, respectively. It is obvious that the winding layer is more prone to fatigue failure compared with the inner liner under random vibration.

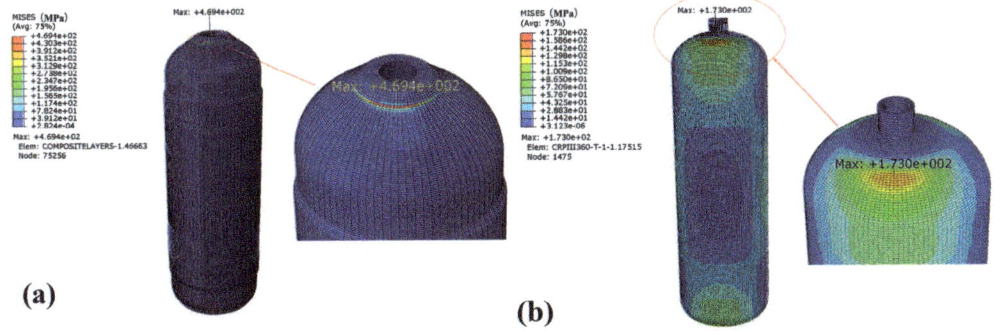

Figure 8. Stress of the (**a**) winding layer and (**b**) inner liner under random vibration.

The root mean square of von Mises (Rmises) stress of the winding layer and the inner liner of the CHSV under random vibration is presented in Figure 9. Rmises stress is the probability distribution of von Mises stress, which is an important research reference in the random vibration study and reflects the magnitude of stress fluctuation. The maximum value of Rmises stress is 86.13 MPa for the winding layer and 62.98 MPa for the liner, respectively, which will be used for subsequent fatigue life prediction.

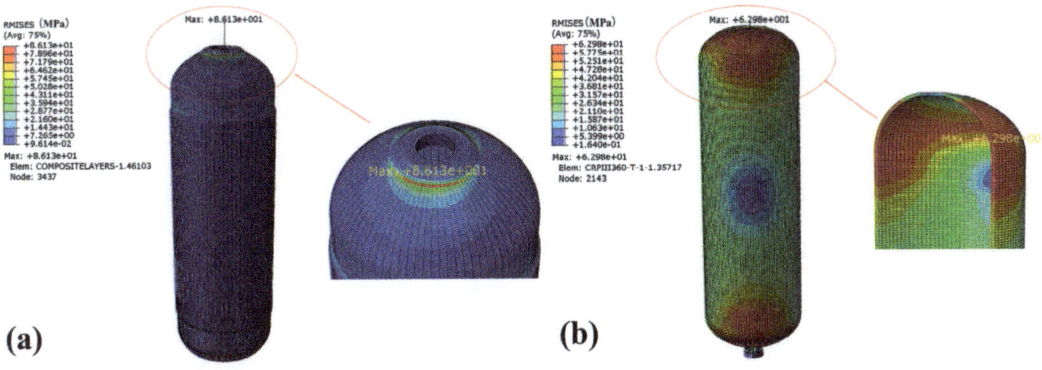

Figure 9. Rmises stress of the (**a**) winding layer and (**b**) inner liner under random vibration.

Figure 10 shows the maximum vibration deformation of the winding layer and the inner liner of the CHSV under random vibration. The maximum values of deformation of both the winding layer and the inner liner located in the center of the vessel are about 0.45 mm and will not produce a large distortion.

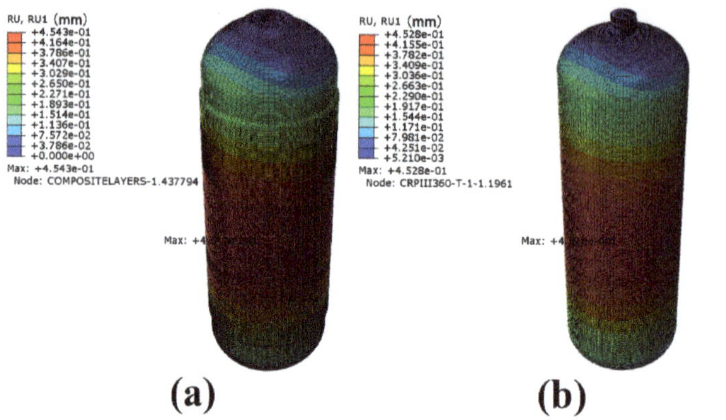

Figure 10. Maximum vibration deformation of (**a**) the winding layer and (**b**) the inner liner of the CHSV under random vibration.

4.3. Fatigue Life Calculations

In the present study, the Steinberg's method is used to predict the fatigue life of CHSV. The Steinberg's method, also known as the three-band method, assumes that the stresses follow a Gaussian distribution and the cumulative damage is a linear combination of damages under different stress levels. The cumulative damage D and the fatigue life n are defined as follows [35]:

$$D = \frac{n_{1\sigma}}{N_{1\sigma}} + \frac{n_{2\sigma}}{N_{2\sigma}} + \frac{n_{3\sigma}}{N_{3\sigma}} \tag{11}$$

$$n = \frac{1}{\frac{0.683}{N_{1\sigma}} + \frac{0.271}{N_{2\sigma}} + \frac{0.043}{N_{3\sigma}}} \tag{12}$$

where $n_{1\sigma}$, $n_{2\sigma}$ and $n_{3\sigma}$ are the actual number of cycles at -1σ~1σ, -2σ~2σ, and -3σ~3σ stress levels, respectively. $N_{1\sigma}$, $N_{2\sigma}$ and $N_{3\sigma}$ are the number of cycles of *S-N* curves at -1σ~1σ, -2σ~2σ, and -3σ~3σ stress levels, respectively. 0.683, 0.271, and 0.043 are the probability of stress distribution at -1σ~1σ, -2σ~2σ, and -3σ~3σ stress levels, respectively.

The fatigue life *S-N* curve of aluminum alloy Al6061-T6 is presented in Figure 11 [36], which is used to predict the life cycle of the CHSV inner liner made of Al6061-T6. In Section 4.2, the maximum values of stresses in the inner liner of CHSV with confidence interval 1σ, 2σ and 3σ are 62.98 MPa, 125.96 MPa, and 188.94 MPa, respectively, corresponding to the numbers of cycles $N_{1\sigma} = 1.04 \times 10^7$, $N_{2\sigma} = 2.31 \times 10^6$, and $N_{3\sigma} = 6.38 \times 10^5$, respectively. According to Equation (12), the number of life cycles of the inner liner can be obtained, as follows:

$$n = \frac{1}{\frac{0.683}{N_{1\sigma}} + \frac{0.271}{N_{2\sigma}} + \frac{0.043}{N_{3\sigma}}} = \frac{1}{\frac{0.683}{1.04\times10^7} + \frac{0.271}{2.31\times10^6} + \frac{0.043}{5.38\times10^5}} = 3.79 \times 10^6 \tag{13}$$

The fatigue life *S-N* curve of CFRP in the winding layer is shown in Figure 12 [37], and the red line shows the fatigue life curve of the specimen reinforced with multiple layers of CFRP. In Section 4.2, the maximum values of stresses of composite layup of the CHSV with confidence interval 1σ, 2σ and 3σ are 86.13 MPa, 172.26 MPa, and 258.39 MPa, respectively, and the corresponding numbers of life cycles are $N_{1\sigma} = 2.14 \times 10^7$, $N_{2\sigma} = 1.22 \times 10^6$, and $N_{3\sigma} = 2.63 \times 10^5$, respectively. The fatigue life can also be calculated according to Equation (12), as follows:

$$n = \frac{1}{\frac{0.683}{N_{1\sigma}} + \frac{0.271}{N_{2\sigma}} + \frac{0.043}{N_{3\sigma}}} = \frac{1}{\frac{0.683}{2.14\times10^7} + \frac{0.271}{8.97\times10^5} + \frac{0.043}{1.92\times10^5}} = 1.79 \times 10^6 \tag{14}$$

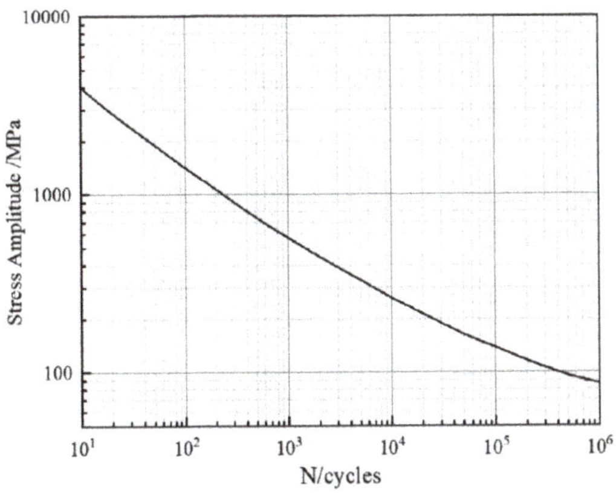

Figure 11. Fatigue life curve of aluminum alloy Al6061-T6 [36].

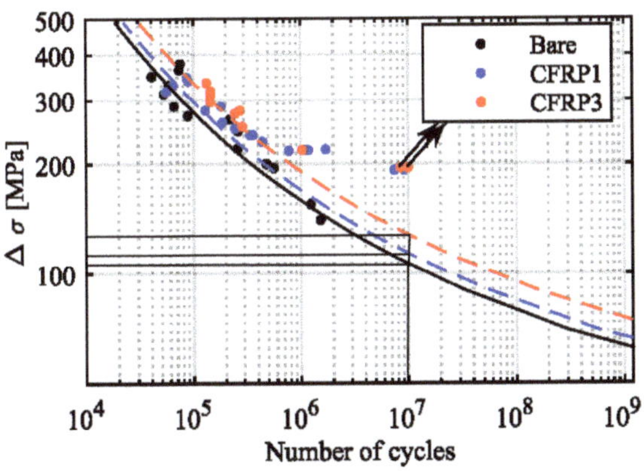

Figure 12. Fatigue life curve of CFRP [37].

In addition, the Rmises stress results of the winding layer and the inner liner obtained in Section 4.2 were then imported into software Fe-safe 2017, in which Dirlik's method was adopted to calculate their fatigue life. Figure 13 shows the results of the fatigue life of the winding layer and the inner liner calculated from the software Fe-safe. The numbers of cycles of the winding layer and the inner liner calculated are $10^{6.221} = 1.66 \times 10^6$ and $10^{6.486} = 3.06 \times 10^6$, respectively. Furthermore, the lowest fatigue life area of the winding layer appears at the CHSV neck, as well as the lowest fatigue life area of the inner liner located at the two ends of CHSV.

The fatigue life of the winding layer and the inner liner calculated by the Steinberg's method and the Dirlik's method are compared in Table 2. The values of the number of life cycles of the winding layer and the inner liner calculated by the Steinberg's method are slightly higher than the results calculated by the Dirlik's method, and the two methods obtain similar prediction results, indicating the accuracy of FEA of fatigue life in the software Fe-safe. Both the values of the number of life cycles of the winding layer and the inner liner are lower than 1×10^7 (high cycle fatigue life standard). In order to further improve the fatigue life of CHSV, the structural optimization of CHSV was then carried out in Section 4.4.

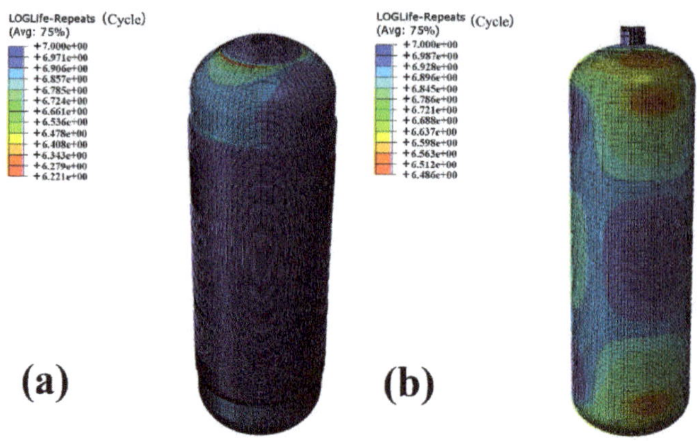

Figure 13. Fatigue life of (**a**) the winding layer and (**b**) the inner liner of the CHSV using software Fe-safe.

Table 2. Random vibration fatigue life estimation results.

Fatigue Life Estimation	Winding Layer	Inner Liner
Number of life cycles (Steinberg method)	1.79×10^6	3.79×10^6
Number of life cycles (Dirlik method)	1.66×10^6	3.06×10^6

4.4. Structural Optimization and Analysis

The most effective and common way to optimize the structure of CHSV is to adjust the spiral winding angle and reduce the thickness of the head section [38]. Reaming is a process to improve the fiber thickness distribution of the head section by changing the helical winding angle, which can make the strength of the head section be distributed more uniformly and improve the bearing capacity of CHSV, without considering the hydrogen embrittlement. Due to the limitation of the curved surface of the head section, the helical winding angle generally ranged from 12° to 70° [39]. Furthermore, the number of layers must be even and consistent based on the characteristics of spiral winding. The above constraint conditions are given as follows [40]:

$$\begin{cases} 13.5^\circ \leq \alpha \leq 70^\circ \\ \alpha_i < \alpha_{i+1} \\ \frac{b}{2R} < \sin\alpha_{i+1} - \sin\alpha_i < \frac{2b}{R} \\ \sum n_i = x \end{cases} \tag{15}$$

where $i = 1, 2, \ldots 6$, x is the number of spiral winding layers, and α the helical winding angle.

In the reaming process, the determining factor of the final thickness distribution of the winding layer is the reaming angle and the corresponding number of winding layers. In the present study, as listed in Table 3, six kinds of winding angles are chosen.

Table 3. Optimized layer laying scheme.

Winding Angle	13.5°	24.1°	36.6°	47.8°	55.5°	90.0°
Number of winding layers	6	2	2	2	2	16

The comparison of winding layers before and after optimization is shown in Figure 14. It can be found that the angle of the winding layer and the stacking of fibers become smoother after the redesign and optimization. The thickness of the fiber layer of the head section is about 13.4 mm before optimization, which is reduced to 7.5 mm after optimization with an obvious decrease of 44%.

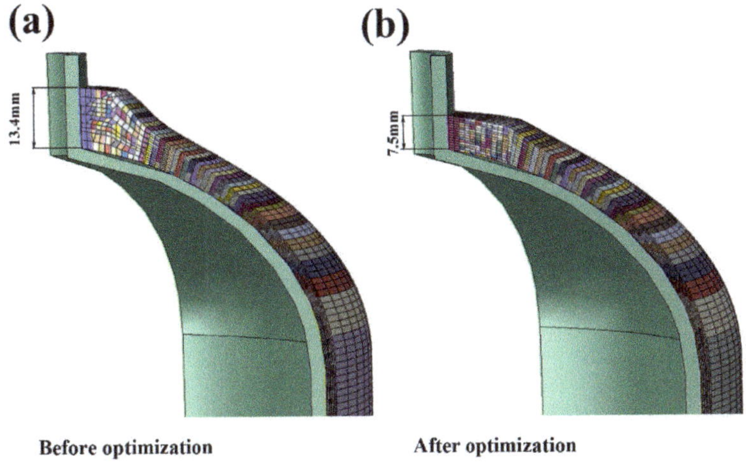

Figure 14. Comparison of winding layers: (**a**) before optimization and (**b**) after optimization.

Figure 15 shows the stress of the winding layer and the inner liner of the CHSV after optimization under the same vibration conditions. Compared with Figure 8, it can be observed that the stresses are still concentrated on the head section of CHSV. However, the maximum value of stress of the winding layer decreases from 469.4 MPa to 156.0 MPa, as well as the maximum value of stress of the inner liner, which decreases from 173.0 MPa to 25.0 MPa. The reduction amplitudes of the maximum stresses of the winding layer and the inner liner reached 66% and 85%, respectively, indicating that the optimization of the winding layer can integrally improve the bearing capacity of CHSV. Similar results of stress reduction after the optimization of the winding layer have been published in the references [41,42].

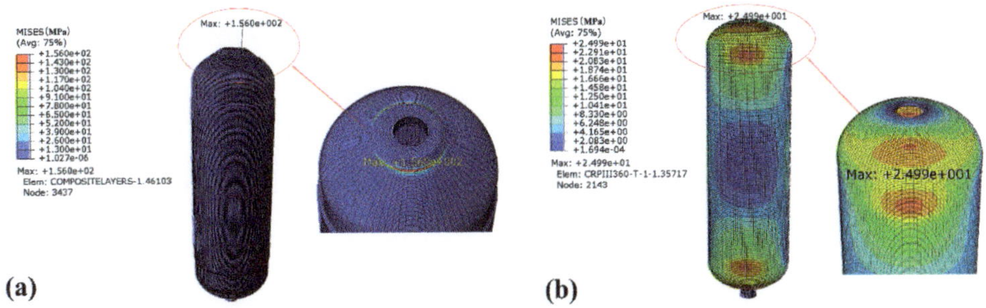

Figure 15. Stress of (**a**) the winding layer and (**b**) the inner liner of the CHSV after optimization.

Fatigue life prediction of optimized CHSV is presented in Figure 16. The number of life cycles of the winding layer and of the inner liner have both been greatly improved, reaching over 1×10^7 (high cycle fatigue life standard). It can be concluded that the optimization of the lay-up angle can effectively improve the safety of the CHSV under random vibration.

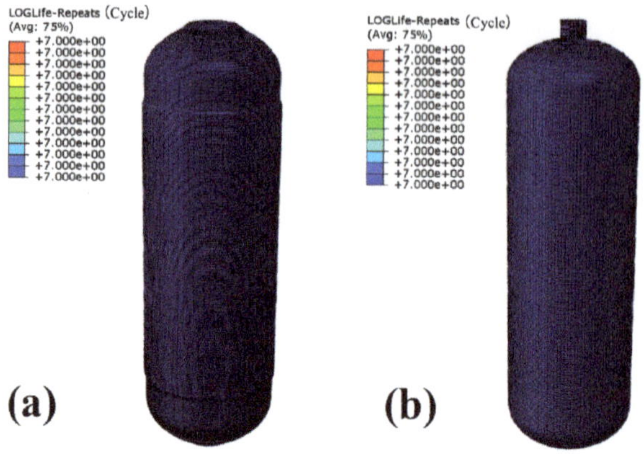

Figure 16. The fatigue life of (**a**) the winding layer and (**b**) the inner liner of the CHSV after optimization.

5. Conclusions

In this paper, the finite element models of CHSV are developed to simulate their dynamic response under the modal testing condition and random vibration condition. Modal testing experiments are conducted to verify the accuracy of numerical simulations. Based on the simulation results of stress under random vibration, the fatigue life of CHSV is subsequently calculated by Steinberg's method and Dirlik's method, respectively. In the end, the optimization of ply parameters of the composite winding layer was conducted to improve the stress distribution and fatigue life of CHSV. Based on the studies, the following conclusions are drawn:

1. The predicted results of the vibration pattern and frequency of CHSV under the free mode and constrained mode conditions show a good agreement with that of the experiment tests. The maximum difference in the value of the vibration frequency for the free and constrained modes of CHSV from the FEA and experiment tests both occur at the first mode and are 8.9% and 8.0%, respectively, verifying the accuracy of the finite element modeling of CHSV;
2. Without the optimization of the winding layer, the maximum values of stress are 469.4 MPa for the winding layer and 173.0 MPa for the liner under random vibration, and both appear at the head section with the thickest fiber accumulation, which indicates that the head section is most likely to occur fatigue failure. Through the optimization of the winding layer, the thickness of the fiber layer of the head section is reduced from 13.4 mm to 7.5 mm, with an obvious decrease of 44%. In addition, the stress of the winding layer and the inner liner after the optimization of the winding layer dropped by 66% and 85%, respectively;
3. There is no obvious difference between the fatigue life of the winding layer and the inner liner calculated by Steinberg's method and Dirlik's method. Without the optimization of the winding layer, the number of life cycles of the winding layer and the inner liner under random vibration conditions obtained from Dirlik's method is about 1.66×10^6 and 3.06×10^6, respectively, while the lowest fatigue life area of the winding layer appears at the CHSV neck and the lowest fatigue life area of the inner liner located in the two ends of CHSV. After the optimization of the winding layer, the numbers of life cycles of the winding layer and the inner liner have been significantly improved, and both reached over 1×10^7 (high cycle fatigue life standard).

The research in this paper provides theoretical guidance for the design, modeling, and optimization of CHSV under random vibration of CHSV, which can be extrapolated to other vessels consisting of the CFRP winding layer and inner liner made of metal materials or polymer materials.

Author Contributions: Methodology, F.F. and J.Z.; Software, M.C.; Writing—original draft, W.W.; Writing—review & editing, X.X.; Supervision, X.L., X.X. and J.Z.; Funding acquisition, X.L. All authors have read and agreed to the published version of the manuscript.

Funding: This research was funded by the Natural Science Foundation of Hubei Province (No. 2024AFB756) and the Key R&D Program of Shandong Province (No. 2024CXGC010215).

Institutional Review Board Statement: Not applicable.

Informed Consent Statement: Not applicable.

Data Availability Statement: The original contributions presented in this study are included in the article. Further inquiries can be directed to the corresponding author.

Conflicts of Interest: The authors declare no conflicts of interest.

References

1. Ratnakar, R.R.; Gupta, N.; Zhang, K.; van Doorne, C.; Fesmire, J.; Dindoruk, B. Hydrogen supply chain and challenges in large-scale LH_2 storage and transportation. *Int. J. Hydrogen Energy* **2021**, *46*, 24149–24168. [CrossRef]
2. Xiao, L.J.; Yao, C.; Shen, W.; Lu, W.G. Layer design of carbon fiber fully wound aluminum cylinder. *J. Mater. Sci.* **2021**, *29*, 32–38.
3. Alves, M.P.; Gul, W. Review on industrial perspectives and challenges on material, manufacturing, design and development of compressed hydrogen storage tanks for the transportation sector. *Energies* **2022**, *15*, 5152. [CrossRef]
4. Bouhala, L.; Karatrantos, A.; Reinhardt, H.; Schramm, N.; Akin, B.; Rauscher, A.; Mauersberger, A. Advancement in the Modeling and design of composite pressure vessels for hydrogen storage: A Comprehensive Review. *J. Compos. Sci.* **2024**, *8*, 339. [CrossRef]
5. Moskvichev, E. Numerical modeling of stress-strain behavior of composite overwrapped pressure vessel. *Procedia Struct. Integr.* **2016**, *2*, 2512–2518. [CrossRef]
6. Kabiri Ataabadi, A.; Ahmadi Jeyed, G. Strength analysis and fatigue life estimation of composite pressure vessel (CPV) based on the residual strain energy (RSE) approach. *Int. J. Comput. Methods Eng. Sci. Mech.* **2020**, *22*, 81–102. [CrossRef]
7. Vafaeesefat, A. Head shape and winding angle optimization of composite pressure vessels based on a multi-level strategy. *Appl. Compos. Mater.* **2007**, *14*, 379–391. [CrossRef]
8. Wang, L.; Wang, B.; Wei, S.; Hong, Y.; Zheng, C.J. Prediction of long-term fatigue life of CFRP composite hydrogen storage vessel based on micromechanics of failure. *Compos. Part B Eng.* **2016**, *97*, 274–281. [CrossRef]
9. Roh, H.S.; Hua, T.Q.; Ahluwalia, R.K. Optimization of carbon fiber usage in Type 4 hydrogen storage tanks for fuel cell automobiles. *Int. J. Hydrogen Energy* **2013**, *38*, 12795–12802. [CrossRef]
10. Cho, S.-M.; Kim, K.-S. Effect of dome curvature on failure mode of type 4 composite pressure vessel. *Int. J. Precis. Eng. Manuf.* **2018**, *19*, 405–410. [CrossRef]
11. Yersak, T.A.; Baker, D.R.; Yanagisawa, Y.; Slavik, S.; Immel, R.; Mack-Gardner, A.; Herrmann, M. Predictive model for depressurization-induced blistering of type IV tank liners for hydrogen storage. *Int. J. Hydrogen Energy* **2017**, *42*, 28910–28917. [CrossRef]
12. Halm, D.; Fouillen, F.; Lainé, E.; Gueguen, M.; Bertheau, D. Composite pressure vessels for hydrogen storage in fire conditions: Fire tests and burst simulation. *Int. J. Hydrogen Energy* **2017**, *42*, 20056–20070. [CrossRef]
13. Rohit, G.; Santosh, M.S.; Kumar, M.N.; Raghavendra, K. Numerical investigation on structural stability and explicit performance of high-pressure hydrogen storage cylinders. *Int. J. Hydrogen Energy* **2023**, *48*, 5565–5575. [CrossRef]
14. Li, J.; Liu, J.; Zhao, B.; Wang, D.; Guo, S.; Song, J.; Li, X. Research on Temperature Rise of Type IV Composite Hydrogen Storage Cylinders in Hydrogen Fast-Filling Process. *Energies* **2023**, *16*, 2918. [CrossRef]
15. Zhou, Y.; Han, B.; Liu, Y. Comparative analysis of impact damage and residual burst pressure of type III and type IV hydrogen storage cylinders. *Int. J. Hydrogen Energy* **2025**, *98*, 370–383. [CrossRef]
16. Hu, Z.; Chen, M.; Pan, B. Simulation and burst validation of 70 MPa type IV hydrogen storage vessel with dome reinforcement. *Int. J. Hydrogen Energy* **2021**, *46*, 23779–23794. [CrossRef]
17. Liu, Y.; Pan, Z.; Li, Q. Comparative analysis of dynamic progressive failure of CFRPs using the scale-span and macroscopic modeling method in drilling. *Eng. Fract. Mech.* **2022**, *268*, 108434. [CrossRef]
18. Liu, H.; Li, B.; Chen, Z. Damage analysis of aluminum alloy gusset joints under cyclic loading based on continuum damage mechanics. *Eng. Struct.* **2021**, *244*, 112729. [CrossRef]

19. Yao, J.; Yin, Y.; Zhang, F.; Qin, Z.; Guo, T. Prediction optimization study of strapping parameters for in-vehicle hydrogen storage cylinders under random vibration based on GA-XGBoost. In Proceedings of the 2024 12th International Conference on Traffic and Logistic Engineering (ICTLE), Macau, China, 23–25 August 2024; pp. 104–109.
20. Huang, J.; Duan, C.; Zhao, M.; Luo, X. Research on frame size of vehicle mounted hydrogen supply system based on parametric design. In *IOP Conference Series: Materials Science and Engineering*; IOP Publishing: Bristol, UK, 2019; Volume 677, p. 032069.
21. Coskun, T.; Sahin, O.S. Modal and random vibration responses of composite overwrapped pressure vessels with various geodesic dome trajectories. *J. Reinf. Plast. Compos.* **2024**. [CrossRef]
22. Ryu, M.; Yun, S.; Kim, B.; Kang, D.; Kim, G.; Lee, H. Numerical analysis of sloshing effects in cryogenic liquefied-hydrogen storage tanks for trains under various vibration conditions. *Cryogenics* **2024**, *143*, 103961. [CrossRef]
23. Wang, X. Study on Strength and Vibration Fatigue Characteristics of a Vehicle-Mounted Liquid Hydrogen Cylinder. *Lanzhou Univ. Technol.* **2022**, 49–68. [CrossRef]
24. Wang, Z.; Wang, Y.; Wang, Q.; Zhang, J. Fuzzy norm method for evaluating random vibration of airborne platform from limited PSD data. *Chin. J. Aeronaut.* **2014**, *27*, 1442–1450. [CrossRef]
25. Zhang, M.; Lv, H.; Kang, H.; Zhou, W.; Zhang, C. A literature review of failure prediction and analysis methods for composite high-pressure hydrogen storage tanks. *Int. J. Hydrogen Energy* **2019**, *44*, 25777–25799. [CrossRef]
26. Decker, M. Vibration fatigue analysis using response spectra. *Int. J. Fatigue* **2021**, *148*, 106192. [CrossRef]
27. Zhang, L.; Cao, S.; Tao, X. Experimental study on interfacial bond behavior between CFRP sheets and steel plates under fatigue loading. *Materials* **2019**, *12*, 377. [CrossRef]
28. Chen, H.; Yang, Y.; Cao, S. Fatigue life prediction of aluminum alloy 6061 based on defects analysis. *Int. J. Fatigue* **2021**, *147*, 106189. [CrossRef]
29. Li, C. Fatigue life estimation based on power spectral density. *Mech. Des. Res.* **2005**, *21*, 6–8.
30. Sofi, T.; Neunkirchen, S.; Schledjewski, R. Path calculation, technology and opportunities in dry fiber winding: A review. *Adv. Manuf. Polym. Compos. Sci.* **2018**, *4*, 57–72. [CrossRef]
31. Zhang, Q.; Liang, S.; Sui, G.; Yang, X. Influence of matrix modulus on the mechanical and interfacial properties of carbon fiber filament wound composites. *R. Soc. Chem.* **2015**, *5*, 25208–25214. [CrossRef]
32. Zhou, J.C.; Hu, C.; Wang, Z.; Ren, Z.; Wang, X.; Mao, K. Study on dynamic characteristics of leaf spring system in vibration screen. *J. Low Freq. Noise Vib. Act. Control* **2021**, *40*, 1818–1832. [CrossRef]
33. Mohamedmeki, Z.M.; Esmail, F.J.; Ajeel, E.A. Fatigue life analysis of laminated elastomeric bearing pad. *Mater. Today Proc.* **2021**, *42*, 2361–2368. [CrossRef]
34. Zhang, Y.P.; Hao, S.Y. Study on the Application of laboratory environmental test methods for military Materiel-GJB 150A—2009. *Pers. Prot. Equip. China* **2017**, *3*, 28–31.
35. García, A.; Sorribes-Palmer, F.; Alonso, G. Application of Steinberg vibration fatigue model for structural verification of space instruments. In *AIP Conference Proceedings*; AIP Publishing: New York, NY, USA, 2018; Volume 1922, p. 100003.
36. Yahr, G. Fatigue design curves for 6061-T6 aluminum. *J. Press. Vessel Technol.* **1997**, *119*, 211–215. [CrossRef]
37. Mohabeddine, A.; Correia, J.; Aires Montenegro, P.; De Jesus, A.; Miguel Castro, J.; Berto, F. Probabilistic S-N curves for CFRP retrofitted steel details. *Int. J. Fatigue* **2021**, *148*, 106205. [CrossRef]
38. Chen, R.X. The analysis and design of filament wound gas cylinder. *J. Solid Rocket Technol.* **2008**, *31*, 625–628, 634.
39. Son, D.; Chang, S. Evaluation of modeling techniques for a type III hydrogen pressure vessel (70MPa) made of an aluminum liner and a thick carbon/epoxy composite for fuel cell vehicles. *Int. J. Hydrogen Energy* **2012**, *37*, 2353–2369. [CrossRef]
40. Jia, Z.X. Structural design and finite element verification of plastic lined composite hydrogen cylinder. *Beijing Univ. Chem. Technol.* **2020**, 57–58. [CrossRef]
41. Sowiński, K. Stress distribution optimization in dished ends of cylindrical pressure vessels. *Thin-Walled Struct.* **2022**, *171*, 108808. [CrossRef]
42. Regassa, Y.; Gari, J.; Lemu, H.G. Composite overwrapped pressure vessel design optimization using numerical method. *J. Compos. Sci.* **2022**, *6*, 229. [CrossRef]

materials

MDPI

Article

Experimental and Numerical Study of the Plasma Arc Melting of Titanium Alloys: Application to the Removal of High Density Inclusions (HDIs)

Jean-Pierre Bellot [1,*], Widad Ayadh [1,2], Jean-Sébastien Kroll-Rabotin [1], Raphaël Marin [2], Jérôme Delfosse [3], Amandine Cardon [3], Alessia Biagi [4] and Stéphane Hans [4]

1 Institut Jean Lamour—UMR CNRS 7198, LabEx DAMAS, Université de Lorraine, 54000 Nancy, France; widad.ayadh@irt-m2p.fr (W.A.); jean-sebastien.kroll-rabotin@univ-lorraine.fr (J.-S.K.-R.)
2 IRT M2P, 4 rue Augustin Fresnel, 57070 Metz, France; raphael.marin@irt-m2p.fr
3 Safran Tech, Rue des Jeunes Bois, 78772 Châteaufort, France; jerome.delfosse@safrangroup.com (J.D.); amandine.cardon@safrangroup.com (A.C.)
4 Aubert & Duval, Site Les Ancizes, BP1, 63770 Les Ancizes CEDEX, France; alessia.biagi@aubertduval.com (A.B.); stephane.hans@aubertduval.com (S.H.)
* Correspondence: jean-pierre.bellot@univ-lorraine.fr

Abstract: Titanium alloys are increasingly used in aeronautical applications, a sector that requires highly controlled materials. In particular, inclusion cleanliness is a necessary and mandatory condition for safe use in aeronautical components. During the production and processing of titanium alloys, inclusions are likely to appear, in particular high-density inclusions (HDIs) originate from refractory metals such as molybdenum or tungsten carbide. Plasma Arc Melting–Cold Hearth Remelting (PAMCHR) is one of the most effective recycling and refining process for titanium alloys. Firstly, this work reports the thermal modeling of the melting of raw materials in the melting crucible and a complete 3D numerical simulation of the thermo-hydrodynamic behavior of the metal flow in the PAMCHR furnace, based on the software Ansys-Fluent CFD V21.1. Simulation results are presented for a 100 kg/h melting test performed in a pilot furnace with a comparison between the measured and calculated pool profiles and residence time distributions that show satisfactory agreements. Additionally, a Lagrangian calculation of particle trajectories in the liquid metal pool is also performed and insemination of HDIs in the pilot furnace has been tested. Both numerical and experimental tests demonstrate the inclusion removal in the melting crucible.

Keywords: titanium alloys; recycling; plasma melting; inclusion; simulation

1. Introduction

Titanium is a metal widely used in the aerospace industry due to its low density compared to steels and superalloys, as well as its excellent mechanical properties, including toughness, corrosion resistance, and ductility. In addition to its classification by the EU as a critical material, the high production cost of titanium—primarily due to the complex and energy-intensive process of manufacturing titanium sponge—makes recycling particularly economically viable [1,2].

Currently, titanium recycling mainly focuses on alloy scraps generated during the manufacturing of aircraft and engine components, such as swarf, cutting chips, and turnings [3]. Beyond its economic benefits, recycling also enhances environmental sustainability by reducing waste, electrical energy consumption, and associated CO_2 emissions.

To either compete with or complement Vacuum Arc Remelting (VAR), an alternative recycling technology utilizing water-cooled, flat-bottomed copper crucibles emerged in the 1990s, offering lower production costs compared to VAR, higher recycling possibility, and better inclusion removal potential [4,5]. Depending on the heat source, this technology is categorized into two processes: Electron Beam Cold Hearth Refining (EBCHR) and Plasma Arc Melting–Cold Hearth Refining (PAMCHR). The EBCHR process employs electron guns in a vacuum environment to ensure proper beam operation, while PAMCHR uses plasma torches under an inert gas atmosphere (Argon or Helium).

For these processes to reach the alloy composition, titanium scraps are melted with some virgin titanium sponge and master alloys in the first melting chamber of the crucible, as illustrated in Figure 1. The liquid metal then flows through a refining crucible before finally reaching a mold casting crucible, where it solidifies into a secondary ingot.

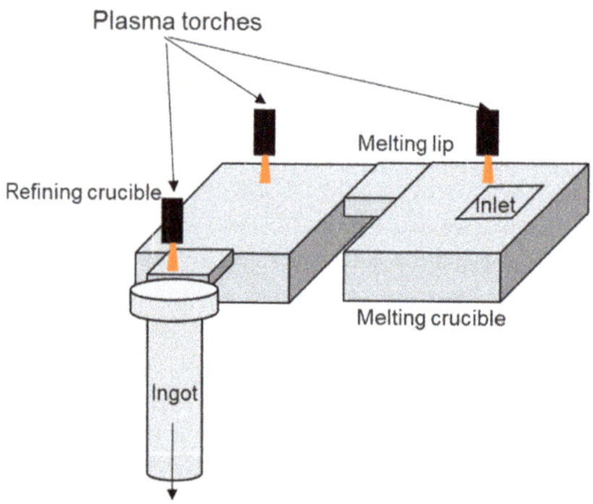

Figure 1. Schematic of remelting processes with water-cooled crucibles and PAMCHR.

Characteristics of the EB process include a relatively quiet molten pool and a quasi-steady distribution of the input heat. But the vacuum environment leads to evaporation losses of the alloy components with a high vapor pressure, such as Al and Cr [6–8]. On the contrary, the very low evaporation is the main advantage of the PAMCHR process against EBCHR, which uses plasma torches as the heat source (operating pressure range between 0.4 and 1 bar).

Despite industrial interest in the PAMCHR process over the past twenty years, the mathematical modeling required to optimize the process remains underdeveloped. A more detailed review of the literature is available at [9]. Modeling began in the late 1990s with the work of Huang et al. [10], including the blowing effect of the plasma jet on the bath surface. Subsequently, in 2013, Xu et al. [11] developed a CFD model based on Ansys-Fluent software V21.1 applied to Ti-Al intermetallic remelting. More recently, in 2018, the Department of Materials Engineering at UBC [12,13] focused their experimental and modeling work on energy transfer from arc plasma to liquid bath.

Inclusion removal is a critical concern for titanium alloys used in aerospace applications. Two types of inclusions can be distinguished based on their density—and consequently—their chemical composition: low-density inclusions (LDIs) and high-density inclusions (HDIs) [14]. Both are exogenous, meaning they form before the remelting process, and their occurrence is extremely rare—some studies estimate only one inclusion per 5000 tons [15]. Due to their diverse origins [16], these particles can vary in size from a few hundred micrometers to several millimeters. If not eliminated during remelting, these inclusions can cause fractures during extended fatigue cycles, such as those experienced by rotating jet engine components. For obvious safety reasons, an alloy free of inclusions is compulsory for engine manufacturing, and remelting processes are specifically designed to remove these two main types of exogenous inclusions. In this paper, we focus exclusively on HDIs, while LDIs will be addressed in a future publication.

HDIs include refractory metals such as tungsten (W), molybdenum (Mo), and tantalum (Ta), along with carbides like tungsten carbide (WC) [17,18]. These defects primarily originate from machining tool fragments. Consequently, recycling machining off-cuts are a major source of HDIs in the recycling materials. Due to their high melting points and slow dissolution rates, HDIs pose a significant challenge. They do not melt and, as noted by Bomberger and Froes [19], typically sink to the bottom of crucibles during remelting, where they dissolve slowly. Yamanaka and Ichihashi [20] investigated the sinking and dissolution behavior of Ta and Mo cylinders in a titanium pool during VAR, using a 100 mm diameter mold. They reported dissolution rates of 0.5 μm/s for Ta and 0.7 μm/s for Mo and show that HDIs might survive the VAR process. In 2010, Ghazal et al. [21] examined the behavior of HDIs (W and Mo) in titanium and titanium alloy (Ti64 and Ti17) baths melted by an electron beam, measuring dissolution rates ranging from 1.6 to 8 μm/s. Finally, Xu et al. [22] employed CFD simulations to demonstrate that HDIs can be removed by trapping them at the bottom of PAM crucibles due to their relative density. The residence time in the liquid bath before entrapment is short, always under four seconds.

The TIARE ("TItanium Aerospace REcycling") project brings together the authors of this paper and aims to support the titanium recycling industry by enhancing the understanding of the recycling process. To this end, IRT-M2P operates a high-capacity pilot furnace equipped with three torches, delivering a total power of 1.2 MW and enabling melting rates of up to 150 kg/h of Ti64. The melting and refining crucibles, along with the ingot mold, are arranged in a U-shape, as in the schematics shown in Figures 1 and 2. As part of the TIARE project, we conducted tests to measure liquid bath profiles, residence time distribution, and seeding with synthetic HDIs. Additionally, a comprehensive PAM3D phenomenological model of the PAMCHR furnace was developed, building on an initial version presented in [9], which modeled heat and momentum transfers between the plasma jet and the liquid surface. PAM3D is currently being further developed at Institut Jean Lamour using the commercial CFD software Ansys-Fluent V21.1. In this new version of the numerical code, both the melting and refining crucibles are modeled, with particular focus on the raw material melting stage. Finally, the dynamic behavior of HDIs is simulated using a Lagrangian approach. The results of the PAM3D model are compared with new experimental data from pilot furnace tests, thus making this paper particularly original.

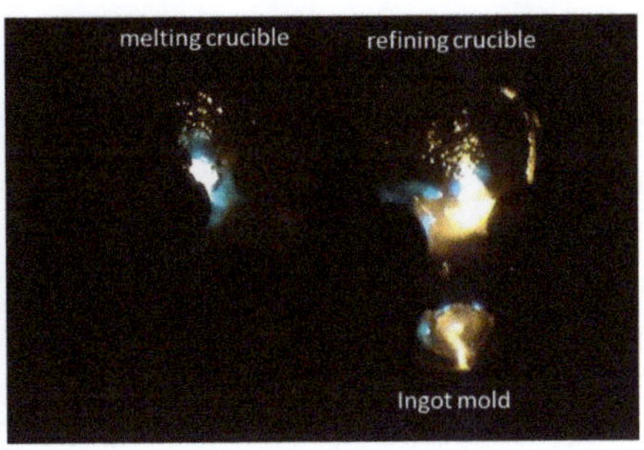

Figure 2. IRT-M2P' pilot furnace (**left**) and top view of the crucibles during processing (**right**).

2. Methods

2.1. *Trials Using the Pilot Furnace*

Three series of tests were carried out using the IRT-M2P pilot furnace, with identical operating conditions, i.e., an electrical power of 358 kW per torch (U = 203 V and I = 1765 A). The scanning pattern of each torch is shown in Figure 3, each torch having a period of 15 s. Torch 1 spends most of its time on the melting zone with the aim of melting the solid input charge.

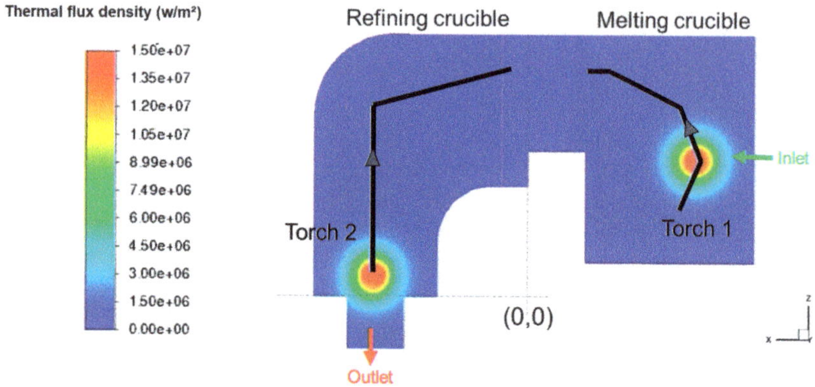

Figure 3. Patterns of the two torches (black lines) and instantaneous heat flux density for a given torch position.

2.1.1. Shape of the Liquid Pool

An initial comparison between the simulation and an experimental test is based on estimating the liquid bath height in the melting and refining crucibles. To facilitate macrographic analysis of the liquid front, the steady-state melting of Ti64 is replaced with TiCp (Commercially Pure), creating a chemical distinction between the liquid bath and the skull. The skull was sectioned at six different locations and in two directions to compare the liquid pool profiles.

2.1.2. RTD Measurement

To experimentally measure the Residence Time Distribution (RTD), a copper marker was added to the titanium load. First, a 37 kg Ti64 bar was melted to establish a quasi-stationary regime. Then, a second Ti64 bar, containing a 127 g copper cube inserted into a drilled hole, was introduced. After melting, the solidified ingot (1233 mm in height and 150 mm in diameter) was cut in half and chemically analyzed through sampling along two longitudinal profiles: one at the center (along the ingot's axis) and the other at the periphery. The copper mass fraction was determined using Inductively Coupled Plasma (ICP) spectrometry.

2.1.3. HDI Insemination

A test was carried out with compacted sponges (called here briquette) seeded with HDIs. In the compact box, 5 mm cubes of Mo, W, and Ta were placed (appearing in yellow—one in the center and one on each side of the box) at four different distances from the front of the box, as seen in the top view of Figure 4. At the outer edge of the box (from 0.60 m or three-quarters of the box), unseeded briquettes were placed. The box was then filled with a second layer of briquettes over the first layer. The total weight of the filled box was 39 kg.

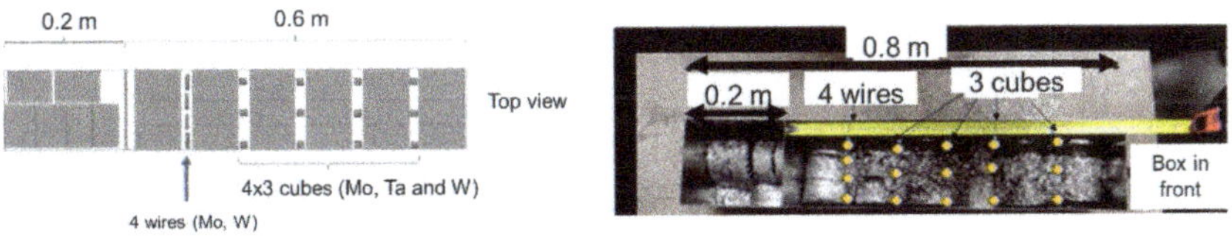

Figure 4. Arrangement of the HDIs in the Ti64 briquette box (**left**), and photograph of the box after deposition of the first briquette layer (**right**).

2.2. Mathematical Modeling

To account for the effects of torch sweeping over the bath, PAM3D models the metal liquid flow in the crucibles in transient and turbulent regimes. To simplify and reduce complexity, we apply the following assumptions:

(i) The surface of the liquid bath is considered to be flat; deformation of the surface under the plasma torch jet is neglected. The torch blast and its action on the bath are taken into account through a shear stress τ_s exerted on the bath surface and distributed as a function of the distance to the point of impact of each torch. The value of this stress has been estimated from a simplified 2D model of the plasma jet [9]. This 2D model and the work provided by Huang et al. [10] demonstrate that the deformation of the free surface can be neglected in a first stage of modeling.

(ii) The effects of the electromagnetic forces are neglected. The electric current transferred between the plasma torch and the liquid bath induces an electromagnetic force. Using a 2D model, a comparison between electromagnetic and natural convection forces showed the predominance of the latter, and consequently the neglect of electromagnetic forces [23].

(iii) The flow of liquid metal is mainly controlled by the movement of the torches. The torches move cyclically along a continuous trajectory with a configurable period. The heat flux and parietal stress distributions due to the torch jets follow a radial distribution around the moving impact point, and their expressions are detailed in [9].

2.2.1. Governing Transport Equations

To avoid making this paper excessively long, the set of governing transport equations (mass, momentum and heat) are not reported here but are extensively described in references [9,23,24]. The reader can find a detailed description of the governing equations in these references. The momentum and heat equations are coupled in particular, but not exclusively, by thermal natural convection modeled through the Boussinesq assumption:

$$\rho = \rho_0 - \beta_T \rho_0 (T - T_0) \tag{1}$$

Because the crucibles are water-cooled, a solid phase called "skull" develops on contact with the cold walls, as shown later. The Kozeny–Carman relation is applied to calculate the permeability K in the mushy zone [25]. The volume fraction of liquid called g_l is expressed as a linear function of the temperature between the solidus (T_{sol}) and liquidus (T_{liq}) of the alloy. Accordingly, the permeability K falls to zero when the temperature reaches the solidus, leading to a zero velocity in the solid phase.

$$K = K_0 \frac{g_l^3}{(1 - g_l)^2} \tag{2}$$

The turbulence in the liquid pool was described with the realizable k-ε model [24], where the production of the turbulent kinetic energy takes into account velocity gradient and buoyancy. This model has become the industry standard for many kinds of engineering problems, especially within the metallurgy industry.

2.2.2. Boundary Conditions

The boundary conditions are summarized in Figure 5, where particular attention has been given to transfers between the plasma and the metal surface [9].

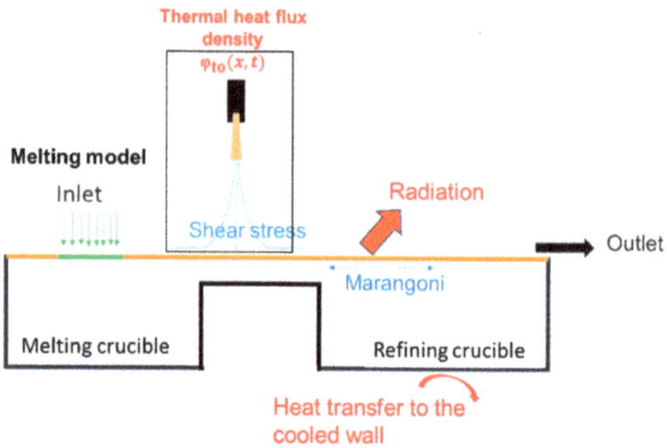

Figure 5. Boundary conditions applied.

The upper surface radiates toward the water-cooled furnace walls. The radiation is expressed by the Stefan–Boltzmann relation using the emissivity ε_m corresponding to the state (solid or liquid) of the metal on the surface:

$$\varphi_{ray}^{up}(T) = \sigma_{sb}\varepsilon_m\left(T^4 - T_{fur}^4\right) \quad (3)$$

with σ_{sb} the Stefan–Boltzmann constant, T the surface temperature, and ε_m = 0.23 for liquid surfaces and 0.60 for solid [26]. T_{fur} is the temperature of the water-cooled furnace walls, assumed to be uniform and equal to 100 °C.

The heat flux density contributed by the torches to the bath surface φ_{to} is calculated from the sum of the electrical $\alpha_{elec}\varphi_{elec}$, convective $\alpha_{conv}\varphi_{conv}$, and radiative $\alpha_{ray}\varphi_{ray}$ contributions of each torch, defined by the following equation:

$$\varphi_{to}(x,t) = \sum_i\left[\alpha_{conv}\varphi_{conv}\left(\left|\vec{x} - \vec{x}_i(t)\right|\right) + \alpha_{elec}\varphi_{elec}\left(\left|\vec{x} - \vec{x}_i(t)\right|\right) + \alpha_{ray}\varphi_{ray}\left(\left|\vec{x} - \vec{x}_i(t)\right|\right)\right]\eta_{to}U_i(t)I_i(t) \quad (4)$$

U_i and I_i are, respectively, the voltage and current at which each torch i is operated, which vary with time, as does the position $x_i(t)$ of each torch's impact point. η_{to} is the torch efficiency, i.e., the ratio between heat power and electrical power. It is estimated to be 0.66 [23]. The shape of the φ_{conv}, φ_{elec}, and φ_{ray} distributions are detailed in [9] and their relative contributions (α coefficients) are noted in Table 1.

Table 1. Coefficients of the relative contributions of the heat provided by the plasma torch [9].

0.66	η_{to}
0.05	$\alpha_{élec}$
0.30	α_{conv}
0.65	α_{ray}

A Fourier condition is applied to model the heat resistance between the solidified alloy and copper wall:

$$\varphi_{wall} = h_{wall}\left(T_p - T_{wall}\right) \quad (5)$$

where T_{wall} is assumed equal to 100 °C. The value h_{wall} = 500 W/m^2/K is then used as the averaged value, which was reported in [27] for the simulation of PAMCHR.

On the liquid surface, a shear stress τ_s is calculated in the region of the arc plasma blast. It varies radially around the impact center of each torch, as follows:

$$\tau_s(r) = f(r)\tau_{max} \tag{6}$$

The function f(r) is zero at the stopping point of the plasma flow, giving a value of zero for the stress. The value of τ_{max} = 10 Pa is retained in the model. The distribution function and the numerical values of the coefficients are given and justified in [9].

Since the surface tension of liquid metal varies with temperature, temperature gradients create surface tension differences, leading to tangential stress—known as the "Marangoni effect". This phenomenon can be described using the following equations:

$$\tau_{yx} = \frac{\partial\sigma}{\partial T}\left(\frac{\partial T}{\partial x}\right)_s \text{ and } \tau_{zx} = \frac{\partial\sigma}{\partial T}\left(\frac{\partial T}{\partial z}\right)_s \tag{7}$$

In the model, the variation of surface tension with temperature ($\partial\sigma/\partial T$) is assumed constant and is provided as input to the simulations.

2.2.3. Melting Model

The melting model is applied to each cell node P of the pre-defined inlet zone (S_{inlet}). The melting rate $\dot{m}_p$ is calculated for each cell P under this feed surface, based on an enthalpy balance, as follows:

$$\left(\varphi_{to}(x_p,t) - \varphi_{ray}^{up}(x_p,t)\right)S_n = \dot{m}_p\left(h_{liq}(T_{sh}) - h_{sol}(T_{ref})\right) \tag{8}$$

The heat flux supplied by the torch minus that lost through radiation is used to heat and melt the solid mass flux, as illustrated in Figure 6. In Equation (8), h_{liq} is defined as the enthalpy of the liquid Ti64 at T_{sh} superheat temperature (inlet value of the model) and h_{sol} is the solid enthalpy at room temperature (25 °C), set to zero as a reference enthalpy in Ansys-Fluent [24]. S_n is the north surface of the cell, see Figure 6.

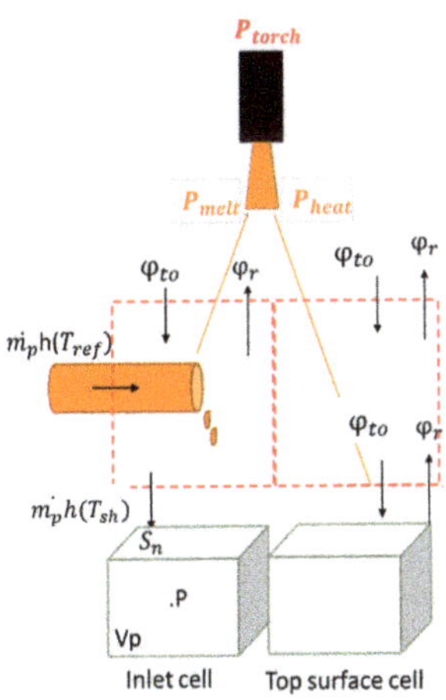

Figure 6. Modeling of the melting rate in each cell of the inlet zone or on the top surface.

For the cells where the net heat flux is negative ($\left(\varphi_{to}(x_p,t) - \varphi_{ray}^{up}(x_p,t)\right) < 0$), no energy is available for melting and $\dot{m}_p = 0$. Otherwise, $\dot{m}_p$ is calculated from Equation (8).

According this melting model, two source terms are introduced for the cells of the inlet zone, a volumetric mass source term:

$$S_{m,P} = \frac{\dot{m}_P}{V_P} \tag{9}$$

and a volumetric momentum source term:

$$S_{mo,P} = -\frac{\dot{m}_P^2}{\rho_0 S_n V_P} \tag{10}$$

where V_P is the volume of cell P. The negative sign in Equation (10) comes from the upward pointing *y*-axis.

2.2.4. Trajectory of High-Density Inclusions (HDIs)

Discrete phase trajectories are simulated using a one-way Lagrangian approach. Particle positions are tracked by integrating their velocity $\vec{u}_p$ over time, which in turn is obtained by integrating their acceleration at every time step. Acceleration is calculated by solving the fundamental equation of dynamics Equation (11) for each particle. In the specific case of HDIs, the lift force can be considered as negligible, so that the equation reduces to:

$$M_p(1 + \frac{\rho}{2\rho_p})\frac{d\vec{u}_p}{dt} = M_p\left(1 - \frac{\rho}{\rho_p}\right)\vec{g} - C_{D,t}\rho\frac{\pi d_p^2}{8}\left\|\vec{u}_{sl}\right\|\left(\vec{u}_p - \vec{u}\right) + \frac{3}{2}M_p\frac{\rho}{\rho_p}\frac{d\vec{u}}{dt} \tag{11}$$

where M_p, d_p, and ρ_p are, respectively, the mass, equivalent diameter, and density of the inclusion. u_{sl} is the sliding velocity between particle and liquid at the particle location. $C_{D,t}$ is a dimensionless drag coefficient. For a particle larger than the Kolmogorov scale η_K, turbulence modifies the flow around the particle, resulting in a modification of the drag coefficient, which according to Brucato et al. [28] is modeled as:

$$C_{D,t} = C_{D,0}\left(1 + 8.67 \cdot 10^{-4}\left(\frac{d_P}{\eta_K}\right)^3\right) \tag{12}$$

$C_{D,0}$ is the drag coefficient of the same moving particle in a stagnant fluid. The latter depends on the particle's Reynolds number according to the correlation [29]:

$$C_{D,0} = \frac{a}{Re_p} + \frac{b}{Re_p^2} + c \tag{13}$$

where the values of a, b, and c depend on Re_p.

Simulation results for the operating conditions of the pilot furnace show that the Kolmogorov scale η_K remains between 30 and 400 μm in the upper part of the bath where turbulence is developed. The $d_p > \eta_K$ condition is therefore verified. Furthermore, the correction (Equation (12)) becomes significant when $8.67 \cdot 10^{-4}\left(\frac{d_p}{\eta_K}\right)^3 > 0.1$, i.e., $d_p \gtrsim 5\,\eta_K$.

On the other hand, the correction (Equation (12)) was established and validated by Brucato et al. for an interval of values represented by the black frame on the left in Figure 7. The simulation results reported in Section 3 present conditions that fall outside this validity interval, which led us to limit the correction to the maximum value of $\frac{(C_{D,t} - C_{D,0})}{C_{D,0}} = 100$, as illustrated in the right side of Figure 7.

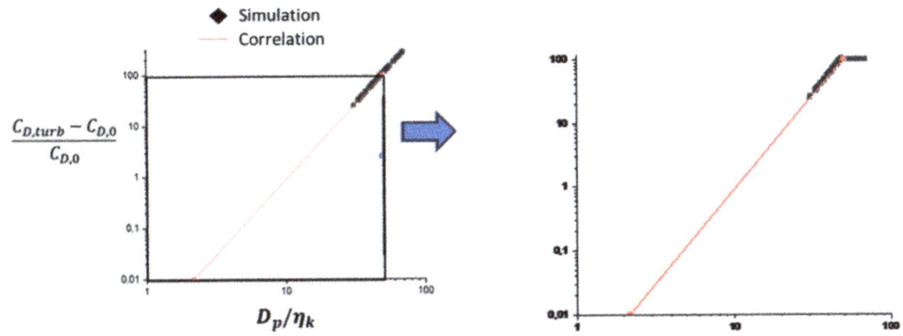

Figure 7. Drag coefficient correction as a function of particle diameter and Kolmogorov scale seen by the particle, given by Brucato [28] (**left**) and applied in PAM3D (**right**).

2.2.5. Numerical Procedure

The 3D domain is meshed with 74,074 hexahedral cells (see Figure 8). The mesh is refined near to the top surface, with a cell height of 0.87 mm. The solution to the set of transport equations is based on the Ansys-Fluent CFD code V21.1 [24], where several user-defined functions (UDF) have been added to account for all the features of the model. Otherwise, the SIMPLEC algorithm is used to solve the Navier–Stokes equations. In addition, the second-order upwind scheme was used for all the *pde* except for momentum, which uses the first-order upwind scheme. Simulations were run on 8 parallel processing cores, leading to the ratio 1 h computation time per 10 s real operating time. The transient numerical simulation with a time-step of 5 ms is carried on until a quasi-stationary state is reached over a torch period (15 s in the actual case). By this time, the thermal and flow conditions in the crucibles have lost the memory of the initial conditions.

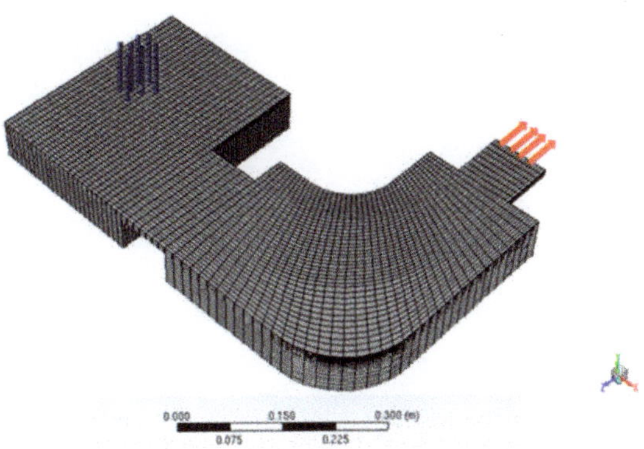

Figure 8. 3D mesh of the IRT'Pilot furnace.

3. Results and Discussion

3.1. Heat Transfer and Fluid Flow

Figure 9 shows temperature maps of the PAMCHR crucible surface at two times (t_1 and t_2), corresponding to different torch locations. At t_1, when the melting torch is above the feed surface, shown as a black outline in Figure 9; overheating is not very pronounced since a significant proportion of the torch power is consumed in melting the loaded titanium and not in heating the bath surface. However, when the melting torch leaves the feed surface at time t_2, a localized zone of higher superheat appears around the torch impact point. Outside the torch impact area, the bath surface temperature averages 1850 °C. We can also see that on the surface of the crucibles, the metal is liquid except on the outlet channel and at the edge of the melting crucible, as shown by the T_{liq} isovalue plotted in black.

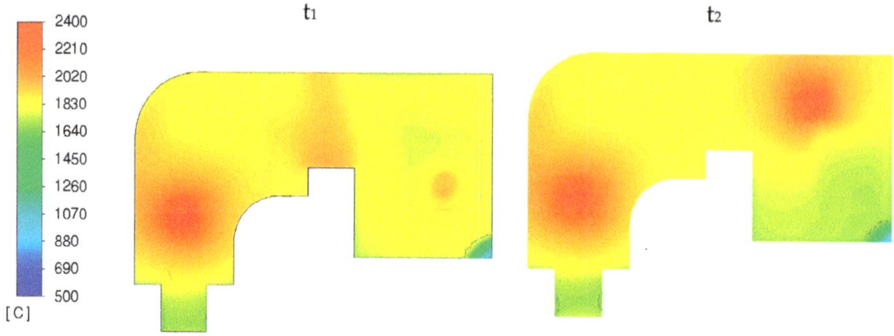

Figure 9. Surface temperature at two different times t_1 (**left**) and t_2 (**right**).

The plasma blast, combined with the thermal Marangoni effect, generates a strong surface flow, with liquid velocities reaching over 20 cm/s, as shown in Figure 10. This kinetic energy is rapidly degraded by turbulence and viscous stress, so that velocities within the liquid pool remain in the 1 to 5 cm/s range. The bath is therefore moderately turbulent in its upper layer (the first 2 cm near the surface), with a dissipation rate $\varepsilon_{max} = 1.5\ m^2/s^3$, and a minimum Kolmogorov length scale of 30 μm.

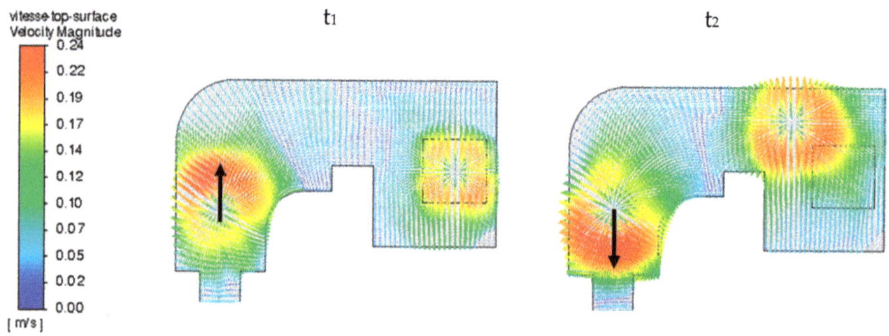

Figure 10. Surface fluid flow at two different times t_1 (**left**) and t_2 (**right**).

It must be noticed that at t_1, the torch on the refining hearth goes downstream, whereas at t_2, this torch moves back upstream, as depicted with the black arrow in Figure 10. Thus, even if the torch blast is modeled without a preferred direction (orthoradial symmetry), a more intense flow upstream from the torch motion is noticed than downstream. This phenomenon can be playfully compared with that of a sheepdog moving through its flock.

3.2. Global Balance

When dealing with computational heat and fluid dynamics, an essential factor to consider is the heat balance reported in Table 2. During a torch period, 39% of the electric power is transferred to heat power (280 kW) on the bath surface: 22% of this power is lost to radiation and 58% lost to the side and bottom walls. Lastly, 18% of the heat power is convected by the liquid metal leaving the refining crucible.

Table 2. Global heat balance over a torch period (15 s).

Electrical Power of the Two Torches (kW)	716.59	
Thermal power provided by the two torches on the crucibles (kW)	281.82	
Accumulation (kW)	0.044	~0.01%
Power lost on crucible walls (kW)	163.73	58%
Power lost by heat radiation (kW)	63.08	22%
Enthalpy heat flux transferred to the ingot * (kW)	52.4	18%
Residual	2.57	0.9%

* Enthalpy reference: Solid Ti64 at room temperature.

3.3. Melting Rate

The melting rate is variable over time, depending on the position of the melting torch relative to the material inlet surface. Figure 11 shows the evolution of the melting rate as a function of time over one melting torch period. It varies from 0 when the torch is outside the feed surface to 140 kg/h when the torch is above this surface, with an average value of 92 kg/h, in excellent agreement with the experimental value (92 kg/h).

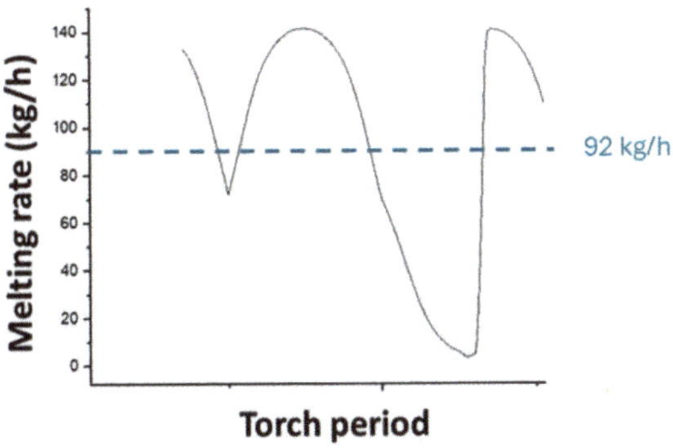

Figure 11. Calculated melting rate over time.

3.4. Liquid Pool Profiles

The skull was sectioned to compare bath depths. Three sections were made in each crucible, as shown in Figure 12. Planes P1, P2, and P3 are located in the melting crucible, while planes P21A, P21B, and P21L are in the refining crucible.

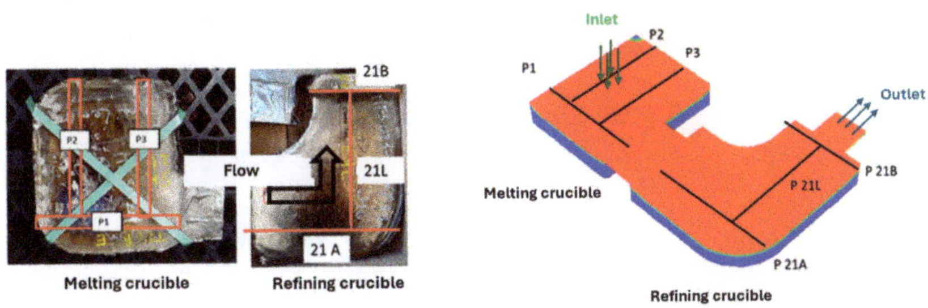

Figure 12. Location of the skull's sections (**left**) and as they appear in the simulation results (**right**).

In Figure 13, the solid fronts in planes P1 and P2 of the melting crucible, obtained both experimentally and through PAM3D simulation, exhibit identical shapes, with similar liquid metal depths of 30 mm and 25 mm, respectively. However, plane P3, located near the transfer channel, shows a greater bath depth in the experiment than in the simulation. This discrepancy can be attributed to the torch trajectory—specifically, its inclination—which is not accounted for in the simulation. This results in a shift in the torch's impact point and an asymmetry in its effect on the bath. Additionally, plane P3 reveals visible differences in the contact quality between the skull and the bottom of the crucible, varying between the left and right sides of the image, which could also contribute to significant variations in bath depth.

In the refining crucible shown in Figure 14, the measured depths in plane P21L align well with the liquid fractions observed in the PAM3D simulation, showing approximately one-third liquid and two-thirds solid. It is much higher than that obtained in a typical electron beam refining hearth, where this ratio is close to 20% [6], and should have an influence on the residence time of inclusions in the liquid pool. The skull shapes in planes P21B and P21A are less accurately captured. In plane P21A, the simulated bath depth is greater than the measured depth. For plane P21B, while the maximum depths in the simulation and experiment are similar, the experimental pool shape is highly asymmetrical—an aspect not captured in the simulation.

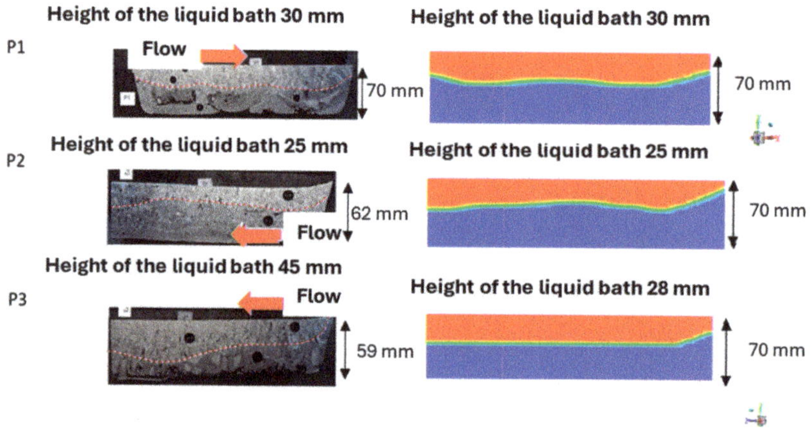

Figure 13. Sections of the melting skull (**left**) and simulated liquid pool (**right**). Flow is given with the red arrow. (Black circular markers locate post-mortem structural or chemical analyses of the alloy). On the right hand side, liquid is in red and solid in blue.

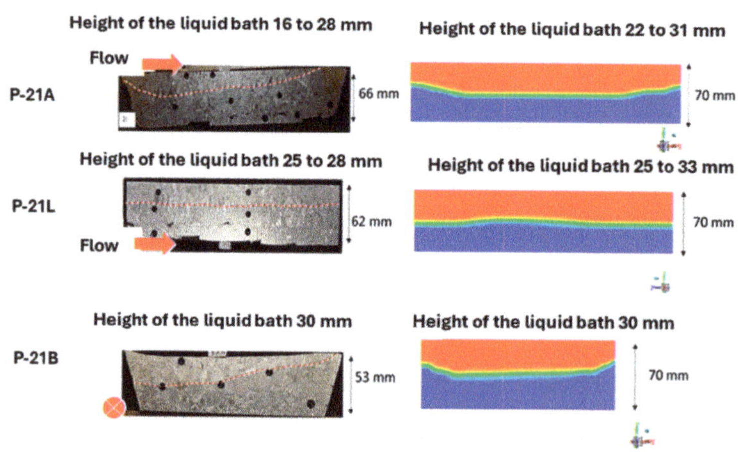

Figure 14. Sections of the refining skull (**left**) and simulated liquid pool (**right**). Flow is given with the red arrow. (Black circular markers locate post-mortem structural or chemical analyses of the alloy). On the right hand side, liquid is in red and solid in blue.

3.5. Residence Time Distribution (RTD)

The simulation of copper transport within the process is achieved by coupling the PAM3D code with a simulation using the PAMELA code, which models the growth and solidification of the final ingot. Further details on the PAMELA software can be found in [30]. The copper mass fraction vs. time signal at the refining crucible outlet, calculated by PAM3D, serves as an input condition for the PAMELA model.

The RTD is reported in Figure 15. The solute first appears at the center of the ingot—slightly lower in height than at the periphery—due to the hollowing effect of the liquid pool. While the simulation overestimates the peak mass fraction value and slightly shifts its position within the ingot, the overall shape of the signal remains consistent, aligning with the behavior of perfectly stirred reactors, as described in [23]. The measurements taken at the top of the ingot (i.e., at the end of the trial) show stronger agreement between the experimental and simulated copper fractions. Despite some discrepancies, the experimental and simulated distributions exhibit similar qualitative trends and are sufficiently close quantitatively, particularly at the end of the test. This confirms that the transport mechanisms and thermo-hydrodynamic phenomena are accurately captured by the numerical model.

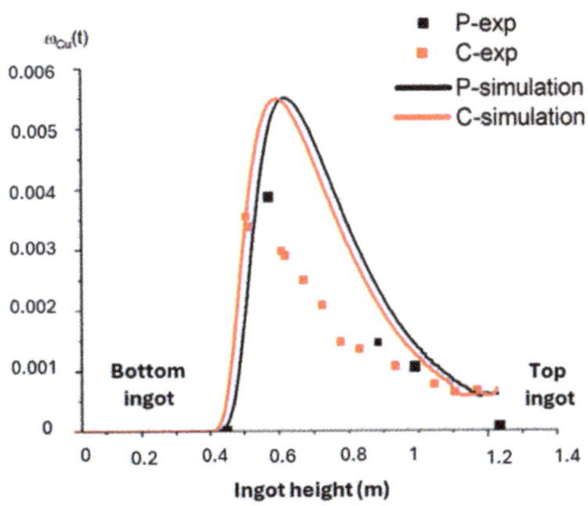

Figure 15. Comparaison of simulated (solid lines) and experimental RTDs (markers). Center of the ingot in red and periphery in black.

3.6. Fate of HDIs

X-ray inspections of the skull were performed using a TC320 RX24 (Cie SGS, Arceuil, France) at a voltage of 270 kV and a current of 12 mA. The resulting X-ray images are shown in Figure 16. The density difference between Ti64 and the HDIs (Mo, W, and Ta) provides sufficient contrast to make the HDIs clearly visible in the images.

All HDIs detected via X-ray are located within the melting crucible, with their trapping positions directly beneath the feeding point—except for one cube, highlighted in yellow in Figure 16, which is slightly displaced from the cluster of other HDIs marked in red. This deviation may be attributed to a slight tilting of the feed box just before melting stopped, causing the cube to be injected slightly farther away.

As expected, due to the significant density difference between the particles and liquid Ti64, the simulated HDIs become trapped by the solid front directly below their injection point. Their residence time in the liquid phase before entrapment is extremely short—less than one second for these 5 mm particles. Calculations varying the HDI size indicate that only particles with an equivalent diameter smaller than 120 μm are transported within the liquid bath. These results agree well with simulations by Xu et al. [11], which also reported residence times of less than one second for particles larger than 1 mm.

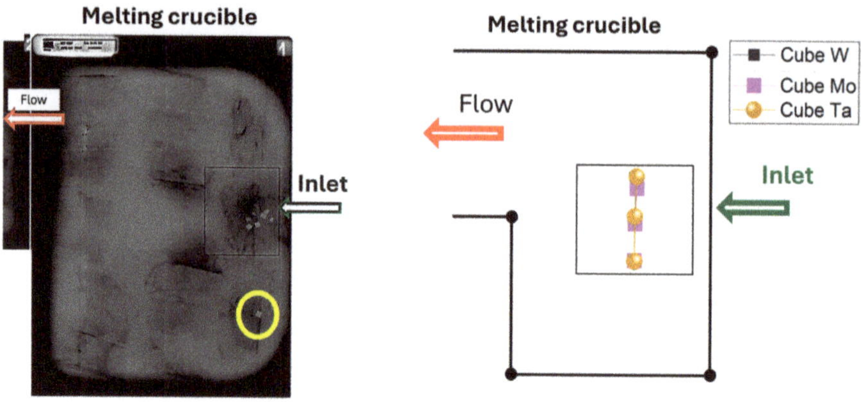

Figure 16. X-ray image of the melting skull (**left**) and calculated location of the trapped HDIs (**right**).

4. Conclusions and Perspectives

The PAM3D code, developed on the Ansys-Fluent platform, is a phenomenological model that describes the thermal–hydraulic behavior of liquid Ti64 in the melting and refining crucibles of the PAMCHR process. The model for the raw material melting stage employs an enthalpy balance to compute the melting rate, considering the plasma torch's power and trajectory within the crucible. As in the reality of the industrial process, the melting speed value changes considerably depending on the position of the melting torch, above the raw material inlet zone or outside it. A global heat balance is calculated over time, showing that over 50% of the thermal energy supplied by the torches is lost in the crucible cooling circuit. The behavior of inclusions is simulated using a Lagrangian approach, specifically applied here to HDI refractory inclusions.

A series of tests were conducted using the IRT-M2P's high-capacity pilot furnace (1.2 MW total electrical power, with a titanium melting rate of up to 150 kg/h) to compare experimental results with numerical simulations. The model accurately predicts the liquid bath profiles, showing a bath depth of approximately one-third of the metal height in the crucible. The residence time distribution of a copper tracer introduced into the raw materials follows a similar trend to the simulation results. Moreover, both the HDI insemination test and the simulation confirm that the process is efficient to remove these particles through sedimentation at the bottom of the melting crucible. This is probably the most important result of this work compared with other processes such as VAR, where efficiency is not totally guaranteed [20,21].

Simulating LDI behavior in the PAMCHR process presents greater complexity, as it requires accounting for dissolution in the Ti64 bath [31,32]. Further experimental and simulation work is planned.

Author Contributions: Methodology, J.-S.K.-R.; Software, W.A.; Investigation, A.C., J.D., S.H. and A.B.; Writing—original draft, J.-P.B.; Supervision, R.M.; Funding acquisition, J.D. and S.H. All authors have read and agreed to the published version of the manuscript.

Funding: This research was funded by National French support PIA (*'Plan d'Investissements d'Avenir'*).

Institutional Review Board Statement: Not applicable.

Informed Consent Statement: Not applicable.

Data Availability Statement: The original contributions presented in this study are included in the article. Further inquiries can be directed to the corresponding author.

Acknowledgments: The authors acknowledge support from the French *"Plan d'Investissements d'Avenir"* (PIA) Program and the National Agency of Research (ANR) within the framework of the TIARE project.

Conflicts of Interest: Authors Alessia Biagi and Stéphane Hans were employed by the company Aubert & Duval. Authors Jérôme Delfosse and Amandine Cardon were employed by the company Safran Tech. The remaining authors declare that the research was conducted in the absence of any commercial or financial relationships that could be construed as a potential conflict of interest.

References

1. Sibum, H. Titanium and Titanium Alloys—From Raw Material to Semi-finished Products. *Adv. Eng. Mater.* **2003**, *5*, 393–398. [CrossRef]
2. Kenjer, A. De L'efficacité des Procédés SMAT et de Choc Laser dans L'amélioration de la Tenue à L'oxydation Haute Température D'alliages de Titane. Ph.D. Thesis, University Bourgogne-Franche Comté France, Besançon, France, 2017.
3. Takanari, O.; Ouchi, T.; Okabe, T. Recent progress in titanium extraction and recycling. *Metall. Mater. Trans. B* **2020**, *51B*, 1315–1328.
4. Sampath, K. The Use of Technical Cost Modeling for Titanium Alloy Process Selection. *JOM* **2005**, *57*, 25–32. [CrossRef]
5. Wood, J.R. Producing Ti-6Al-4V plate from single-melt EBCHM ingot. *JOM* **2002**, *54*, 56–58. [CrossRef]
6. Bellot, J.-P.; Hess, E.; Ablitzer, D. Aluminum volatilization and inclusion removal in the electron beam cold hearth melting of Ti alloys. *Metall. Trans. B* **2000**, *31B*, 845–854. [CrossRef]
7. Reimer, C.; Jourdan, J.; Bellot, J.P. Ti-17 volatilization under vacuum: Investigation of the activity coefficients at 1900 °C. *Calphad* **2017**, *58*, 122–127. [CrossRef]
8. Truong, V.-D.; Hyun, Y.-T.; Woon, J.; Lee, W.; Yoon, J. Numerical simulation of the effects of scanning strategies on the aluminum evaporation of titanium alloy in the electron beam cold hearth melting process. *Materials* **2022**, *15*, 820. [CrossRef]
9. Bellot, J.-P.; Decultot, L.; Jardy, A.; Hans, H.; Doridot, E.; Delfosse, J.; McDonald, N. Numerical simulation of the plasma arc melting cold hearth refining process (PAMCHR). *Metall. Mater. Trans. B* **2020**, *51B*, 1329–1338. [CrossRef]
10. Huang, X.; Chou, J.; Tilly, D.; Yu, K. Computer Simulation of the Refining Hearth in a Plasma Arc Melting Process. In Proceedings of the Liquid Metal Processing and Casting Conferences, Santa Fe, NM, USA, 5–8 October 1997.
11. Xu, X.; Chang, H.; Kou, H.; Yang, Z.; Li, J. Simulation of Temperature Field Distribution in Melting TiAl Alloy by PAM Process. *Int. J. Innov. Tech. Explor. Eng.* **2013**, *2*, 1123–1133.
12. Ji, S.; Duan, J.; Yao, L.; Maijer, D.; Cockcroft, S.; Fiore, D.; Tripp, D. Quantification of the heat transfer during the plasma arc re-melting of titanium alloys. *Int. J. Heat Mass Transfer.* **2018**, *119*, 271–281. [CrossRef]
13. Yao, L.; Maijer, D.; Cockcroft, S.; Fiore, L.; Tripp, D. Quantification of heat transfer phenomena within the melt pool during the plasma arc re-melting of titanium alloys. *Int. J. Heat Mass Transf.* **2018**, *126*, 1123–1133. [CrossRef]
14. Blackburn, M.; Malley, D. Plasma arc melting of titanium alloys. *Mater. Des.* **1993**, *14*, 19–27. [CrossRef]
15. Woodfield, A.; Lemaitre, G. Aerospace Titanium Alloy Melt Process Quality Improvements. In Proceedings of the MATEC Web of Conferences, Piešťany, Slovakia, 7–9 October 2020.
16. Shamblen, C.; Hunter, G. Titanium Base Alloys Clean Melt Process Development. In the General Electric Aircraft Engines Series Report R89AERB141. 1989.
17. Garrat, P.; Neal, D. The origin and dissolution of nitride and dense metal inclusions in vacuum arc remelting of titanium alloys. In Proceedings of the Liquid Metal Processing and Casting Conferences, Santa Fe, NM, USA, 11–14 September 1994.
18. Cen, M.; Liu, Y.; Zhang, H.; Li, Y. Inclusions in melting process of titanium and titanium alloys. *China Foundry* **2019**, *16*, 223–231. [CrossRef]
19. Bomberger, H.; Froes, F. The melting of titanium. *JOM* **1984**, *36*, 39–47. [CrossRef]
20. Yamanaka, A.; Ichihashi, H. Dissolution of refractory elements to titanium alloy in var. *ISIJ Int.* **1992**, *32*, 600–606. [CrossRef]
21. Ghazal, G.; Chapelle, P.; Jardy, A.; Jourdan, J.; Millet, Y. Dissolution of high density inclusions in titanium alloys. *ISIJ Int.* **2012**, *52*, 1–9. [CrossRef]
22. Xu, X.; Kou, H.; Yang, Z.; Li, J. Simulation of HDIs behaviors in the plasma arc cold hearth melting process. *Adv. Mat. Res.* **2014**, *893*, 672–678. [CrossRef]
23. Décultot, L. Etude et Modélisation du Procédé de Refusion par Plasma d'arc en Creuset Froid (PAMCHR) D'alliages de Titane Pour des Applications Aéronautiques. Ph.D. Thesis, Université de Lorraine, France, Nancy, 2021.
24. Ansys, A. *Fluent Solver Theory Guide*; ANSYS Inc.: Canonsburg, PA, USA, 2021.
25. Murakami, K.; Okamoto, T. Fluid flow in the mushy zone composed of granular grains. *Acta Metall.* **1984**, *32*, 1741–1744. [CrossRef]

26. Choi, W.; Jourdan, J.; Matveichev, A.; Jardy, J.; Bellot, J.-P. Kinetics of Evaporation of Alloying Elements under Vacuum. *High Temp. Mater. Process.* **2017**, *36*, 815–823. [CrossRef]
27. Huang, X.; Chou, J.; Pang, Y. Modeling of Plasma Arc Cold Hearth Melting and Refining of Ti Alloys. In Proceedings of the Liquid Metal Processing and Casting, Santa Fe, NM, USA, 21–24 February 1999.
28. Brucato, A.; Grifasi, F.; Montante, G. Particle drag coefficients in turbulent fluids. *Chem. Eng. Sci.* **1998**, *53*, 3295–3315. [CrossRef]
29. Morsi, S.; Alexander, A. An investigation of particle trajectories in two-phase flow systems. *J. Fluid Mech.* **1972**, *55*, 193–208. [CrossRef]
30. Doridot, E.; Hans, S.; Jardy, A.; Bellot, J.-P. Industrial applications of modelling tools to simulate the PAMCHR casting and VAR process for Ti64. In Proceedings of the 14th Word Conference on Titanium, Nantes, France, 10–14 June 2019.
31. Bellot, J.-P.; Foster, B.; Hans, S.; Hess, E.; Ablitzer, D.; Mitchell, A. Dissolution of hard-alpha inclusions in liquid titanium alloys. *Met. Mat. Trans. B* **1997**, *28B*, 1001–1010. [CrossRef]
32. Mizukami, H.; Kitaura, T.; Shirai, Y. Nitriding behavior of titanium sponge studied using nitrogen gas and dissolution behavior of a titanium nitride sponge in titanium alloy melt. *ISIJ Int.* **2019**, *59*, 104–112. [CrossRef]

materials

MDPI

Article

Theoretical and Experimental Research on the Short-Range Structure in Gallium Melts Based on the Wulff Cluster Model

Chun Wang, Minghao Hua, Luyao Wang, Shenglong Wang, Jinlong Liu, Rong Liu, Xuelei Tian and Xiaohang Lin *

Key Laboratory for Liquid-Solid Structural Evolution and Processing of Materials, Ministry of Education, Shandong University, Jinan 250061, China; 202314148@mail.sdu.edu.cn (C.W.); huaminghao@outlook.com (M.H.); 202334253@mail.sdu.edu.cn (L.W.); 202234232@mail.sdu.edu.cn (S.W.); 202214130@mail.sdu.edu.cn (J.L.); 202414167@mail.sdu.edu.cn (R.L.); tianxuelei@sdu.edu.cn (X.T.)

* Correspondence: lxh12345@sdu.edu.cn; Tel.: +86-531-88392727

Abstract: In this paper, the short-range ordering structures of Ga melts has been investigated using the Wulff cluster model (WCM). The structures with a Wulff shape outside and crystal symmetry inside have been derived as the equivalent system to describe the short-range-order (SRO) distribution of the Ga melts. It is observed that the simulated HTXRD patterns of the Ga WCM are in excellent agreement with the experimental data at various temperatures (523 K, 623 K, and 723 K). This agreement includes first and second peak positions, widths, and relative intensities of patterns, particularly at temperatures significantly above the melting point. A minor deviation in the second peak position has been observed at 523 K, attributed to the starting of the pre-nucleation stage. These findings demonstrate that the WCM can effectively describe the SRO structure in melt systems exhibiting a certain extent of covalency.

Keywords: gallium melts; Wulff cluster model; short-range ordering; high-temperature XRD; DFT; surface energy

1. Introduction

In recent years, gallium and its alloys have played an important role in many fields such as semiconductor devices; solar cells; catalysis [1–8], especially for the manufacture of microwave communication and microwave integration; infrared optics and infrared detector devices; integrated circuits; and light-emitting diodes, etc. [9,10]. To enhance the quality and improve the gallium production manufacturing techniques, a deep understanding of the short-range structures of Ga melts is necessary. This is the key to the properties of the melt and subsequent products.

To investigate the melt structure, experimental methods such as HTXRD [11], X-ray absorption fine structure (EXAFS) [12,13], neutron scattering [14], and so on, are widely used. Intensity patterns and pair distribution functions (PDFs) have been intensively studied through intensity correction and Fourier transformation [15–17], although the fundamental physical picture of the SRO of the melts is obscured by the high-temperature and liquid experimental conditions. In other words, the physical definition of the liquid state remains incomplete. In this case, simplified physical models such as the hard sphere model, microcrystalline model, and cluster model [18–21] have been built to help researchers understand the experimental phenomena. However, determining the atomic-scale structure of metallic melts is still challenging. This is intertwined with several unresolved issues, such as the non-Arrhenius changes in viscosity and the amorphous SRO structures.

In prior research, a thermodynamic statistical model, namely WCM, was built by our group with the aim of describing the SRO of the melts. In this model, the most probable microstructure is regarded as an equivalent system of SRO in melts within a state of thermodynamic equilibrium. The most probable microstructure is given by the Wulff shape within the corresponding crystalline structure. Meanwhile, the size of the microstructure is derived directly from the experimental data obtained via HTXRD. This model is successfully applied to melts of pure metals [22], binary homogeneous alloys [23], eutectic alloys [24], and alloys with intermetallic compounds [25]. All of the above systems have a common feature, namely, that the bonding within them is a metallic bond. The covalent bond system (like Ga) is a kind of system that has never been studied in this way. The widespread use of semiconductors makes the study of the melt structure within a covalent bond system very necessary because it is still unclear.

In this article, the melt SRO of a system with covalent bonds (Ga) is investigated using the Wulff cluster model under thermodynamic equilibrium. The suitability of WCM in the system with covalent bonds has been proven by HTXRD experiments.

2. Methods

2.1. Experimental Methods

The pure gallium specimen employed in this research consisted of 15 g of pure gallium with a purity of 99.99%. Prior to the measurement, we ensured that there was no oxide on the specimen. Subsequently, the molten specimen was analyzed using HTXRD(Jinan, China) with a Mo target serving as the X-ray source. The sample was superheated at 773 K for 50 min within an yttria crucible, and then gradually cooled to the target temperatures (723 K, 623 K, and 523 K). Before the liquid X-ray diffraction patterns were measured, the samples were maintained at the respective temperatures for 30 min. The specific parameters were as follows: Generator Current was set at 40 mA, Generator Voltage was 45 KV, and the scanning angle (2θ) was from 10° to 80° with 0.5° step length. The exposure time for each step was 30 s. After atomic scattering correction and a nonlinear transformation of the intensity curve, the structure factor curve S(Q) can be derived using the following formula [25]:

$$S(Q) = \frac{I(\theta)}{Nf(Q)^2} \tag{1}$$

Here, $f(Q)$ represents the atomic form factor, which is obtained from the International Tables for Crystallography [26]; N is defined as the number of atoms engaged in diffraction; Q, which is a nonlinear scaling of θ, is the magnitude of inverted lattice vector, computed as $Q = 4\pi sin\theta/\lambda$, where λ designates the X-ray wavelength employed in the HTXRD measurement. The PDF $\rho(r)$ is employed to depict the number density of atoms on a spherical shell centered around an atom with a radius r from the center. It constitutes a fundamental approach for probing the distribution of atoms within the molten state. $S(Q)$ could be described by the formula as follows [25]:

$$S(Q) = 1 + \int_0^{\infty} 4\pi r^2[\rho(r) - \rho_0]\frac{sin(Qr)}{Qr}dr \tag{2}$$

Here, ρ_0 represents the average density determined by macroscopic density and atomic mass. Although the utilization of certain auxiliary functions and mathematical approximation methods emerged, the PDF $\rho(r)$ could be acquired by sinusoidal transformation.

Experimentally [27,28], the correlation radius R_c, according to Formula (3), is regarded as the size of the equivalent structure [25].

$$\left|\frac{\rho(r)}{\rho_0} - 1\right| < 0.02 \quad (r > R_c) \tag{3}$$

2.2. Simulation Methods

In the present article, the Vienna ab initio Simulation Package 6.3 (VASP 6.3) is utilized for first principle calculation with the aim of obtaining the surface energy γ. The generalized gradient approximation (GGA) exchange-correlation potential and the Perdew–Burke–Ernzerhof (PBE) exchange-correlation function are chosen [29,30]. The projector augmented wave (PAW) pseudopotential potentials constructed by Kresse and Joubert is used [31,32]. After conducting convergence tests, an energy cutoff of 500 eV is selected. The Monkhorst–Pack scheme with $7 \times 7 \times 1$ k-points is selected to ensure a satisfactory convergence of the energy for slab modeling. Concurrently, the alternative Monkhorst–Pack scheme with $7 \times 7 \times 7$ k-points is deemed an appropriate selection for pure gallium bulk modeling (as will be detailed subsequently). The relaxation of most stable structures continues until the residual forces of the total structure are less than 0.01 eV/Å. In prior research, it has been established that the relative proportions of distinct faces in the Wulff shape, as determined by interface energy, closely approximate those computed via γ [23,33,34]. Therefore, it is reasonable to use the computed γ instead of interface energy to describe the structure of the Wulff model.

The influence of temperature on various surfaces has been meticulously investigated. It has been proved that a finite temperature hardly influences the γ. Specifically, when compared with the surface energy at 0 K, the discrepancy in surface energy remains less than 0.5% even at 800 K [21–24]. Consequently, γ at 0 K can be adopted for the construction of the Wulff shape.

For the purpose of calculating the surface energy of different crystallographic plane families, the double-faced slab model, which is shown in Figure 1, is employed. In this model, specific atoms are extracted from a bulk supercell, thereby creating a vacuum layer and simultaneously exposing the desired crystal plane. One lattice vector c of the supercell is arranged to be perpendicular to the exposed surface, whereas the other two lattice vectors are configured to be parallel to it. The atoms located in the vicinity of the center of the slab model, which are considered the atoms belonging to the bulk phase, are fixed (ordinarily approximately 60%). In contrast, the outermost atoms, which are representative of the surface atoms, are permitted to relax. The surface energy, denoted as γ, is determined by [35]:

$$\gamma = \frac{1}{2A}(E_{salb} - NE_{bulk}) \tag{4}$$

In Equation (4), A designates the area of the crystallographic surface that is exposed within the model. E_{salb} represents the total energy of the system, E_{bulk} corresponds to the total energy of the primitive cell in the bulk phase (bulk model), while N signifies the quantity of primitive cells.

To enable a comparison with the experiment data, the X-ray diffraction (XRD) pattern is simulated. This is achieved by inversely solving the Laue equation using the aforementioned cluster. The angle step length is configured to be 0.05°, and the full width at half maxima (FWHM) is configured to be smaller than the step length, thereby facilitating the acquisition of peak position. In the output data, the peaks with intensities less than 10% of the maximum peak intensity are excluded. Subsequently, the residual peaks are expanded in accordance with the Gaussian peak function to mimic the small-size effect and the significant lattice distortion of the atomic cluster structure within a melt. A temperature background function is incorporated to simulate the alterations in the X-ray intensity distribution that result from temperature variations. The ultimate broadening formula is as follows [36]:

$$I(2\theta) = \sum_{i=1}^{n} \left\{ I_i \left[\frac{P_1}{\sqrt{2\pi} \cdot \sqrt{a_1^2 + b^2}} e^{-\frac{(2\theta - 2\theta_i)^2}{2(a_1^2 + b^2)}} + \frac{P_2}{\sqrt{2\pi} \cdot \sqrt{a_2^2 + b^2}} e^{-\frac{(2\theta - 2\theta_i)^2}{2(a_2^2 + b^2)}} \right] \right\} \quad (5)$$

where I_i is associated with the intensity of simulation peaks; P_1 denotes the proportion of the number of inner atoms to the total amount of atoms within the WCM; P_2 represents the ratio of outer atoms; a_1 and a_2 relate to the peak widening induced by the inner and outer lattice distortions, respectively; and b is related to the peak widening caused by fine crystallization. The parameters were ascertained based on experimental observations and were suitably adjusted within an appropriate range.

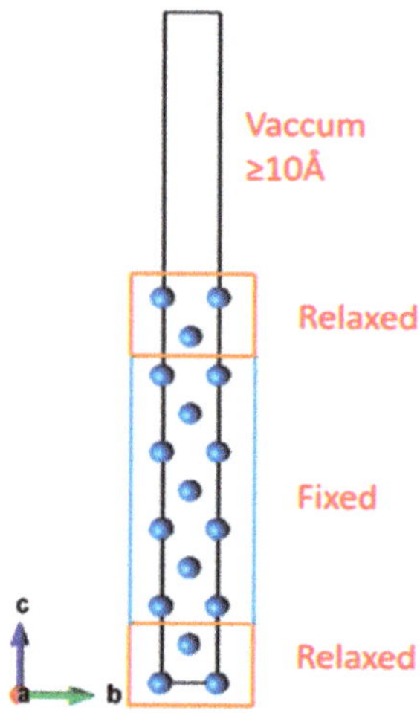

Figure 1. Demonstration of the double-faced slab model.

3. Results and Discussion

3.1. HTXRD Results of Pure Gallium

The HTXRD patterns of the pure Ga samples at temperatures of 523 K, 623 K, and 723 K are displayed in Figure 2a. The discrepancies among the intensity curves under these three temperatures are rather subtle, and the positions of the most prominent peaks do not exhibit significant alterations. The peaks are attributed to the SRO structures (clusters) within the melts, while the background height is a consequence of the free atoms scattering (disordered atoms). It can be found that the strongest peak of each curve shifts to the right with a very small amplitude as the temperature shifts from high to low, which appears at 16.4° (723 K), 16.6° (623 K), and 17° (523 K). At the same time, the position of the secondary peak moves right slightly with the reduction in temperature, which appears near 32° and has a wider peak shape compared with the first peak. In addition, the first peak is about four times stronger than the second one. To obtain the pair distribution function, the temperature background ought to be removed from the experimental data; then, the data are normalized and transformed into S(Q) through Formula (1). S(Q) undergoes sinusoidal transformation to $\rho(r)$ by Formula (2). Subsequently, it is divided by the PDF ρ_0 to obtain $g(r)$, shown in Figure 2b. From Formula (3), the average correlation radii of the melts at 523 K, 623 K, and 723 K are 11.11 Å, 10.50 Å, and 8.91 Å, respectively. These radii exhibit a tendency to increase with the decrease in temperature, which is in consonance with the prior anticipation.

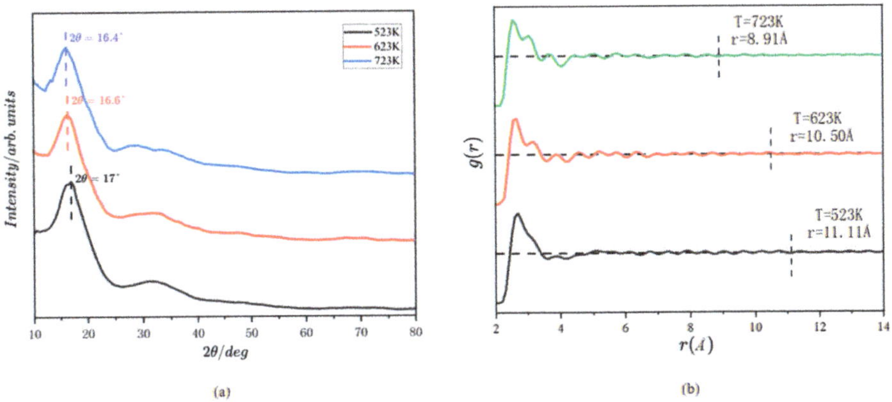

Figure 2. (**a**) XRD intensity pattern and (**b**) PDF of the pure gallium melt at temperatures of 523 K, 623 K, and 723 K.

3.2. Basic Characteristics of Gallium

Gallium is a kind of special metal that has certain extent covalency in the crystal state. All of the computational structures are founded upon α-Ga, which adopts an orthorhombic crystal system with four atoms residing within the primitive cell. Each atom is in proximity to seven neighboring atoms, thereby giving rise to a greatly anisotropic atomic environment, and it is the single crystal phase of gallium that can exist chronically under normal pressure and temperature [37–42]. The specific structure is shown in Figure 3.

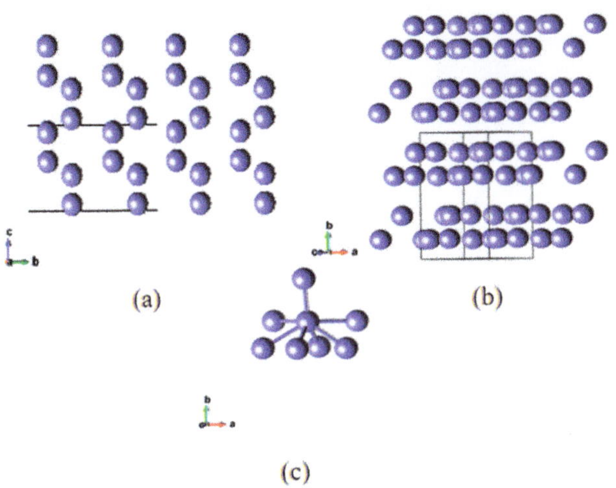

Figure 3. (**a**,**b**) Pure gallium bulk structure in two perspectives. (**c**) Gallium atom coordination relationship.

Compared with the experiment data [43], the gallium crystal lattice parameters shown in Table 1 have a maximum error of only 3%, which indicates that the chosen model and parameters are reasonable.

Table 1. Lattice parameter (Å) of the Ga bulk.

Parameter	A (Å)	B (Å)	C (Å)	Nearest Neighboring (Å)	Second Nearest Neighboring (Å)
Experiment	4.51	4.49	7.65	2.44	2.71–2.79
Calculation	4.63	4.55	7.77	2.51	2.73–2.77

In Table 1, A, B, and C represent lattice constants. Each Ga atom has one very close neighboring atom and six atoms only a little further away, which is shown in the table.

The charge density distributions of Cu and Ga are shown in Figure 4. The red part in the color axis denotes the region of high charge density, and the blue and black parts in the color axis represent the region of low charge density. Generally speaking, the valence electrons of typical metals (like Cu) exhibit nonlocalized characteristics. In Figure 4a, the blue area indicates the uniform charge density in the Cu bulk. Unlike typical metals, the valence electrons are more localized between the Ga atoms (area in the figure consisting of a hexagonal ring marked in the figure), which shows in the red color in Figure 4b. The other areas have relatively low charge densities between the rings of six Ga atoms (the area surrounded by the black line). It is obvious that the bonds exhibit the properties of covalent bonds in the Ga crystal.

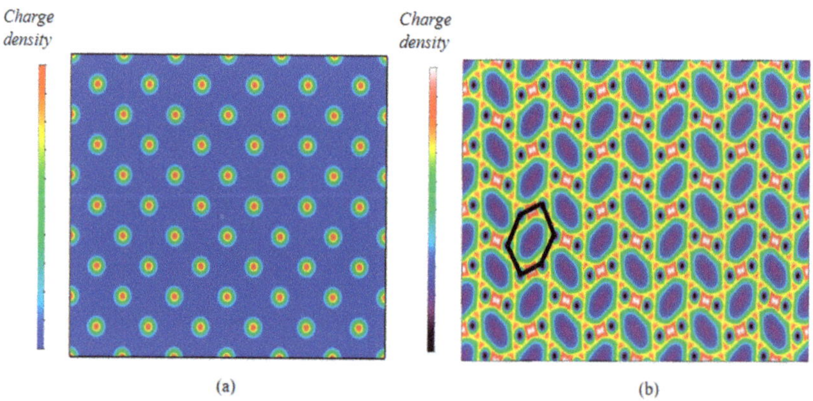

Figure 4. (**a**) The charge density distribution of the Cu bulk from the view of Cu (100) direction. (**b**) The charge density distribution of the Ga bulk from the view of Ga (100) direction.

3.3. *Wulff Cluster Model of Ga*

The HTXRD experiments mentioned above were repeated serval times at each temperature with extremely consistent results, which obviously demonstrates the melts are a thermodynamic equilibrium system. On the one hand, according to the Wulff construction, the most probable shape should have the lowest total surface Gibbs free energy, determined by the interface energies [35]. On the other hand, the electron distribution, which should obey Fermi–Dirac distribution, determines many properties of the crystals, especially the point group symmetry. The electron distribution hardly changes significantly until the temperature reaches the Fermi degeneracy temperature (FDT). Considering that the FDT typically exceeds in the order of magnitude of 10^4 K, the temperature of metallic crystals and melts (generally in the order of magnitude between 10^2 K and 10^3 K) are considerably low. As a result, the trend in point group symmetry hardly changes between these two phases. Hence, the structures with a Wulff shape outside and crystal symmetry inside have been derived as the equal system for characterizing the SRO distribution within the microstructure of the melts.

Although the Wulff cluster model that has the lowest total surface Gibbs free energy on the cluster's surface could describe all kinds of metals in principle, the applicability in the crystal with the properties of covalent bonds still needs investigation. To determine the Wulff shape of Ga, surface energies of Ga surfaces are calculated using the two-side slab model mentioned above. The result of the calculation determines which crystal planes will be exposed on the cluster surface. The established slab model is shown in Figure 5. Low Miller index crystal planes with low surface energies are selected to establish the slab modeling and compute the γ. The specific data of γ are displayed in Table 2. All of the data are relatively reasonable compared with other DFT calculation results [44].

Figure 5. Top and side views of the (**a**) Ga (101) surface, (**b**) Ga (120) surface, (**c**) Ga (201) surface, (**d**) Ga (100) surface, (**e**) Ga (001) surface, (**f**) Ga (111) surface, (**g**) Ga (102) surface. The area surrounded by the orange line in the figure is the unit of the surface. All surfaces that are exposed on the Wulff shape are shown in the figure.

Table 2. Surface energy calculation of Ga cluster.

Crystal Plane	Surface Energy γ (J/m^2)	Percentage (%)	Crystal Plane	Surface Energy γ (J/m^2)	Percentage (%)
Ga (001)	0.591	21.670	Ga (102)	0.643	0.690
Ga (100)	0.477	30.960	Ga (201)	0.590	0.770
Ga (101)	0.618	1.380	Ga (211)	0.605	0.000
Ga (111)	0.592	18.640	Ga (310)	0.554	0.000
Ga (120)	0.607	25.960	Ga (010)	0.847	0.000

As shown in Table 2, the Ga (100) plane, among all crystalline planes calculated, has the minimum γ (0.477 J/m^2). During the calculation process, we found that the distance between the nearest neighboring atoms on the (100) surface decreased and dimerization occurred, which is the main reason for the lowest γ. For the low-index surfaces of Ga that are calculated, the trend of surface energy $\gamma_{100} < \gamma_{001} < \gamma_{101}$ is apparent. These results indicate that the density of each crystal plane also has an impact on surface energy, and the (100) plane is precisely the most close-packed plane of Ga. Meanwhile, the relatively low-index crystal planes are prone to being exposed on the Wulff cluster. In addition, the crystal planes can be exposed more on the cluster when the surface energy γ of the plane decreases to a certain extent compared with the other planes. On the basis of the modeling and above-mentioned calculational results, the shape of WCM that is depicted according to the calculated γ data is displayed in Figure 6a. For the Ga cluster, the (100) surface's ratio of corresponding exposed area is the largest (about 30.96%). The shape of the WCM of the Ga cluster is similar to a spindle shape.

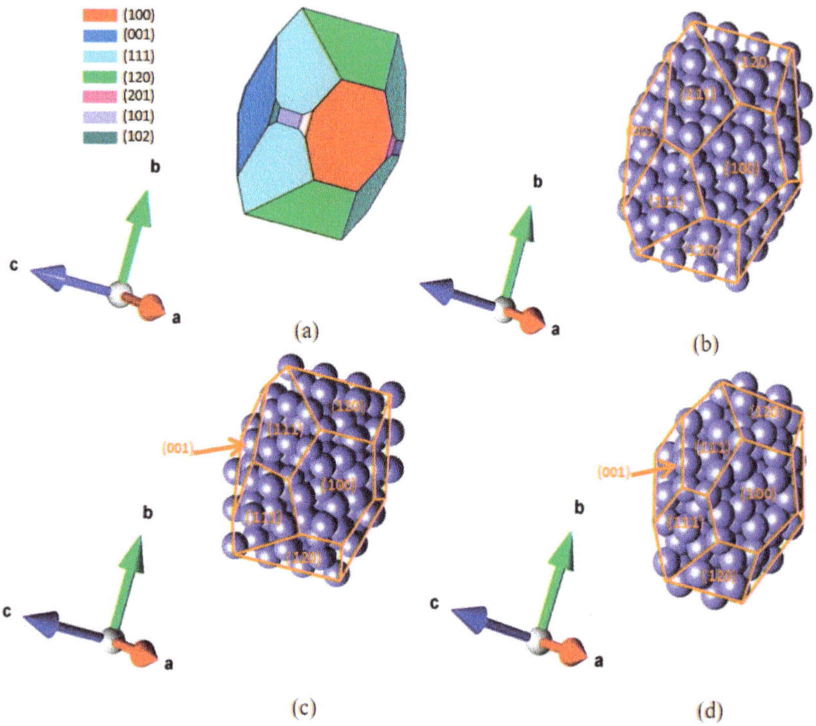

Figure 6. (**a**) The schematic of the Ga Wulff shape. The atomic equivalent structure of Ga melts at (**b**) 523 K, (**c**) 623 K, and (**d**) 723 K.

To acquire the Wulff cluster, the average correlation radii and other related data mentioned above are given by the HTXRD experiments, and the atomic equivalent structures of the Ga melts are displayed in Figure 6b–d. It can be observed that for the Ga Wulff atomic microstructure, the majority of the calculated faces exist. Due to the small cluster size, relatively high index faces with relatively low surface energies, such as the Ga (210) and Ga (310) surfaces, cannot be exposed on the Wulff cluster.

To check the accuracy of WCM within the melts featuring covalent bonds, it is necessary to conduct a direct comparison between the calculated XRD results and the experiment data shown in Figure 7. The blue perpendicular lines denote the diffraction patterns of the atomic structures above at 0 K, the red curve is the experimental XRD results, and the black one is the calculated broadening XRD curve. Not only the positions but also the comparative intensities of the two peaks of the simulated XRD curves agree with the experimental data quite well, especially at high temperatures (623 K and 723 K). Note that a slight deviation of the position of the second peak at 523 K takes place. Such phenomena happen at low temperatures near the melting point and have been observed in almost all metallic alloy systems [19,21,22]. It is usually caused by the beginning of the pre-nucleation stage. Determining the short-range structure of gallium melts serves as a foundation for further research into gallium-based liquid alloys, providing us the opportunity to understand the inner interactions within these materials. This gives us the chance to design the composition and processing techniques of gallium-based liquid metals with an industrial product-oriented approach, which will significantly promote the development of industries such as lubrication and flexible electrodes.

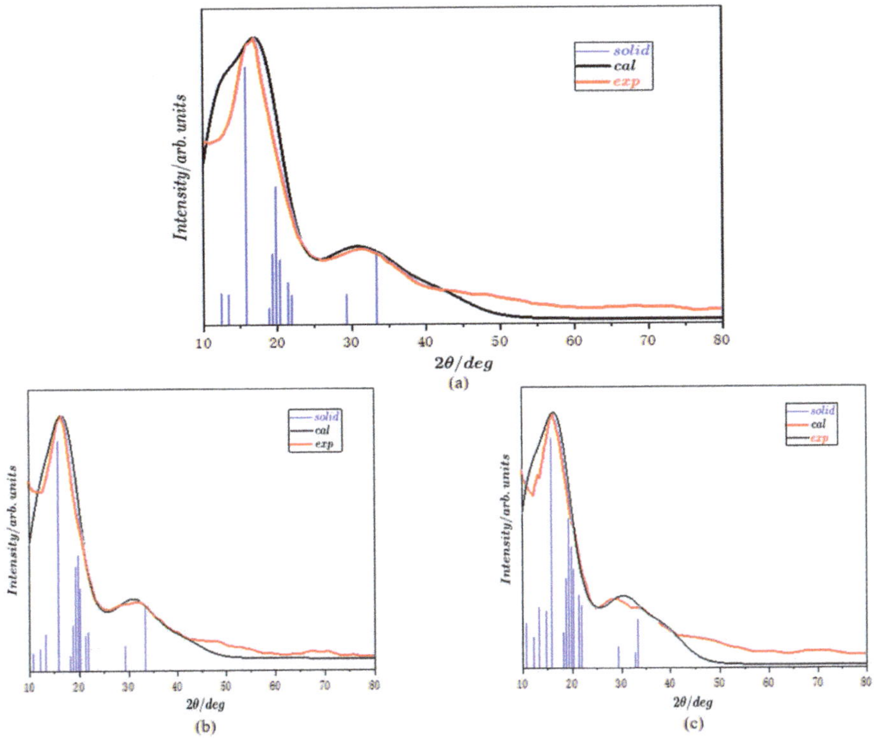

Figure 7. The XRD simulation results compared with the experimental data at (**a**) 523 K, (**b**) 623 K, and (**c**) 723 K.

4. Conclusions

In this paper, the SRO structures of Ga melts has been investigated by the WCM, which has been proven to successfully describe the melt structure of pure metals, homogenous alloys, eutectic alloys, and alloys with intermetallic compounds. Structures with Wulff shapes outside and crystallographic symmetry inside have been derived as the equivalent system to describe the SRO distribution of the Ga melts. The XRD curves of Ga melts at different temperatures (523 K, 623 K, 723 K) were obtained in the experiment, and the $g(r)$ and the correlative radius r (11.11 Å, 10.50 Å, 8.91 Å) were obtained accordingly. The γ data of different crystal planes of Ga crystals were acquired in the DFT calculation, and the γ of the (100) plane was the lowest (0.477 J/m^2). Based on the above results, the atomic equivalent structures at different temperatures have been presented. Through comparison with the experimental outcomes of HTXRD, it has been discovered that overlaying the HTXRD pattern of the gallium WCM enables the simulated HTXRD pattern to be in excellent agreement with the experimental data at the designated temperatures. This includes aspects such as the peak position, width, and relative intensity, particularly at temperatures significantly above the melting point. A tiny deviation of the second peak position at 523 K occurred, which was caused by the beginning of the pre-nucleation stage.

In this case, the WCM can describe the SRO structure of melt systems with a certain extent of covalency, such as Ga.

Author Contributions: Methodology, C.W.; formal analysis, M.H. and C.W.; investigation, C.W. and R.L.; data curation, C.W., X.L. and L.W.; writing—original draft preparation, C.W. and J.L.; writing—review and editing, X.L.; visualization, X.T. and S.W.; supervision, X.L.; project administration, X.T. and X.L.; funding acquisition, X.L. All authors have read and agreed to the published version of the manuscript.

Funding: This research was funded by the basic research priorities program of the National Natural Science Foundation of Shandong, China (Project No. ZR2021ZD22). The APC was funded by Shandong University.

Institutional Review Board Statement: Not applicable.

Informed Consent Statement: Not applicable.

Data Availability Statement: The data that support the findings of this study are available within the article.

Conflicts of Interest: The authors declare no conflicts of interest.

References

1. Andrearczyk, T.; Levchenko, K.; Sadowski, J.; Gas, K.; Avdonin, A.; Wróbel, J.; Figielski, T.; Sawicki, M.; Wosinski, T. Impact of Bismuth Incorporation into (Ga, Mn) As Dilute Ferromagnetic Semiconductor on Its Magnetic Properties and Magnetoresistance. *Materials* **2023**, *16*, 788. [CrossRef] [PubMed]
2. Ciocărlie, L.; Negrea, A.; Ciopec, M.; Duteanu, N.; Negrea, P.; Svera, P.; Ianăşi, C. Adsorption Performances and Mechanisms of $MgFe_2O_4$ Spinel Toward Gallium (III) from Aqueous Solution. *Materials* **2024**, *17*, 5740. [CrossRef]
3. Levchenko, I.; Kryvyi, S.; Kamińska, E.; Smalc-Koziorowska, J.; Grzanka, S.; Kacperski, J.; Nowak, G.; Kret, S.; Marona, L.; Perlin, P. Pros and Cons of $(NH_4)_2S$ Solution Treatment of p-GaN/Metallization Interface: Perspectives for Laser Diode. *Materials* **2024**, *17*, 4520. [CrossRef]
4. Jackson, P.; Hariskos, D.; Lotter, E.; Paetel, S.; Wuerz, R.; Menner, R.; Wischmann, W.; Powalla, M. New world record efficiency for Cu (In, Ga)Se_2 thin-film solar cells beyond 20%. *Prog. Photovolt. Res. Appl.* **2011**, *19*, 894–897. [CrossRef]
5. El-Bahy, Z.M.; Ohnishi, R.; Ichikawa, M. Hydrolytic, Hydrolytic decomposition of CF_4 over alumina-based binary metal oxide catalysts: High catalytic activity of gallia-alumina catalyst. *Catal. Today* **2004**, *90*, 283–290. [CrossRef]
6. Sun, X.; Li, H. Recent progress of Ga-based liquid metals in catalysis. *RSC Adv.* **2022**, *12*, 24946–24957. [CrossRef]
7. Yuan, J.; Du, J.; Hou, Y.; Chen, F.; Li, Q. Carbon and Silicon Impurity Defects in GaN: Simulating Single-Photon Emitters by First Principles. *Materials* **2024**, *17*, 3788. [CrossRef]
8. Liu, Y.; Huang, R.; Lin, T.; Dang, J.; Huang, H.; Shi, J.; Chen, S. Preparation of Sn-Doped Ga2O3 Thin Films and MSM Ultraviolet Detectors Using Magnetron Co-Sputtering. *Materials* **2024**, *17*, 3227. [CrossRef] [PubMed]
9. Moskalyk, R. Gallium: The backbone of the electronics industry. *Miner. Eng.* **2003**, *16*, 921–929. [CrossRef]
10. Mašek, J.; Kudrnovsky, J.; Máca, F.; Sinova, J.; MacDonald, A.H.; Campion, R.P.; Gallagher, B.L.; Jungwirth, T. Mn-doped Ga(As,P) and (Al,Ga)As ferromagnetic semiconductors: Electronic structure calculations. *Phys. Rev. B* **2007**, *75*, 045202. [CrossRef]
11. Kovalev, D.Y.; Vadchenko, S.G.; Rogachev, A.S. In Situ High-Temperature X-ray Diffraction Study of the Thermal Stability of the $Co_{0.22}Cr_{0.23}Fe_{0.29}Ni_{0.20}Ti_{0.06}$ High-Entropy Alloy Produced by High-Energy Ball Milling. *Adv. Eng. Mater.* **2024**, *26*, 2401107. [CrossRef]
12. Reale, P.; Pinto, V.; Cayado, P.; Celentano, G.; Armenio, A.A.; Rufoloni, A.; Santoni, A.; D'acapito, F. Evidence of Gd substitution for Y in YBCO films with Gd excess. *J. Alloys Compd.* **2024**, *983*, 173921. [CrossRef]
13. Feng, Z.C.; Liu, J.; Xie, D.; Nafisa, M.T.; Zhang, C.; Wan, L.; Jiang, B.; Lin, H.-H.; Qiu, Z.-R.; Lu, W.; et al. Optical, Structural, and Synchrotron X-ray Absorption Studies for GaN Thin Films Grown on Si by Molecular Beam Epitaxy. *Materials* **2024**, *17*, 2921. [CrossRef] [PubMed]
14. Melnikov, N.; Reser, B.; Paradezhenko, G. Spin-density correlations in the dynamic spin-fluctuation theory: Comparison with polarized neutron scattering experiments. *J. Magn. Magn. Mater.* **2016**, *411*, 133–139. [CrossRef]
15. Tostmann, H.; DiMasi, E.; Pershan, P.S.; Ocko, B.M.; Shpyrko, O.G.; Deutsch, M. Microscopic surface structure of liquid alkali metals. *Phys. Rev. B* **2000**, *61*, 7284–7287. [CrossRef]
16. Wen, P.; Demaske, B.; Spearot, D.E.; Phillpot, S.R.; Tao, G. Effect of the initial temperature on the shock response of $Cu_{50}Zr_{50}$ bulk metallic glass by molecular dynamics simulation. *J. Appl. Phys.* **2021**, *129*, 165103. [CrossRef]

17. Lou, H.; Wang, X.; Cao, Q.; Zhang, D.; Zhang, J.; Hu, T.; Mao, H.-K.; Jiang, J.-Z. Negative expansions of interatomic distances in metallic melts. *Proc. Natl. Acad. Sci. USA* **2013**, *110*, 10068–10072. [CrossRef]
18. Bernal, J.D. A Geometrical Approach to the Structure of Liquids. *Nature* **1959**, *183*, 141–147. [CrossRef]
19. Bernal, J.D.; Finney, J.L. Random Packing of Spheres in Non-rigid Containers. *Nature* **1967**, *214*, 265–266. [CrossRef]
20. Huang, L.; Wang, C.Z.; Ho, K.M. Structure and dynamics of liquid $Ni_{36}Zr_{64}$ by ab initio molecular dynamics. *Phys. Rev. B* **2011**, *83*, 184103. [CrossRef]
21. Monkhorst, H.J.; Pack, J.D. Special points for Brillouin-zone integrations. *Phys. Rev. B* **1976**, *13*, 5188–5192. [CrossRef]
22. Song, L.; Tian, X.; Yang, Y.; Qin, J.; Li, H.; Lin, X. Probing the Microstructure in Pure Al & Cu Melts: Theory Meets Experiment. *Front. Chem.* **2020**, *8*, 607.
23. Song, L.; Tian, X.; Shao, A.; Li, L.; Zhang, Y.; Li, H.; Lin, X. The structure of metallic melts in binary homogenous alloys: A thermodynamical understanding from the Wulff cluster model. *Phys. Chem. Chem. Phys.* **2020**, *22*, 23237–23245. [CrossRef]
24. Song, L.; Tian, X.; Shao, A.; Hua, M.; Li, L.; Li, H.; Lin, X. The structure of metallic melts in eutectic alloys based on the Wulff cluster model: Theory meets experiment. *Phys. Chem. Chem. Phys.* **2021**, *23*, 3606–3614. [CrossRef] [PubMed]
25. Shao, A.; Tian, X.; Song, L.; Hua, M.; Xue, J.; Wu, S.; Lin, X. Theoretical and experimental study of the microstructure of a metallic melt in an $In_{50}Bi_{50}$ alloy based on the Wulff cluster model. *Phys. Chem. Chem. Phys.* **2022**, *24*, 10798–10806. [CrossRef]
26. Brown, P.; Fox, A.; Maslen, E.; Keefe, M.O.; Willis, B. *Intensity of Diffracted Intensities*; Springer: Dordrecht, The Netherlands, 2004; pp. 554–595.
27. Zhao, Y. Structural Evolution and Cluster Behavior in the Solidification Process of Binary Alloy's Melts. Ph.D. Thesis, Shandong University, Jinan, China, 2007.
28. Yang, Z.X.; Geng, H.R.; Zhang, N. Anomalous Change of the Viscosity and the Structure of Sn Melt with Temperature. *Adv. Mater. Res.* **2010**, *123–125*, 931–934. [CrossRef]
29. Hohenberg, P.; Kohn, W. Inhomogeneous Electron Gas. *Phys. Rev.* **1964**, *136*, B864. [CrossRef]
30. Kohn, W.; Sham, L.J. Self-Consistent Equations Including Exchange and Correlation Effects. *Phys. Rev.* **1965**, *140*, A1133. [CrossRef]
31. Kresse, G.; Joubert, D. From ultrasoft pseudopotentials to the projector augmented-wave method. *Phys. Rev. B* **1999**, *59*, 1758–1775. [CrossRef]
32. Blöchl, P.E. Projector augmented-wave method. *Phys. Rev. B* **1994**, *50*, 17953–17979. [CrossRef]
33. Mathew, K.; Kolluru, V.S.C.; Mula, S.; Steinmann, S.N.; Hennig, R.G. Implicit self-consistent electrolyte model in plane-wave density-functional theory. *J. Chem. Phys.* **2019**, *151*, 234101. [CrossRef]
34. Mathew, K.; Sundararaman, R.; Letchworth-Weaver, K.; Arias, T.A.; Hennig, R.G. Implicit solvation model for density-functional study of nanocrystal surfaces and reaction pathways. *J. Chem. Phys.* **2014**, *140*, 084106. [CrossRef] [PubMed]
35. Gross, A. *Theoretical Surface Science: A Microscopic Perspective*, 2nd ed.; Springer: Berlin, Germany, 2009.
36. Shao, A. Mechanistic Study and Qualitative Physical Modeling of the Viscosity Jump Phenomenon in Metal Melts. Ph.D. Thesis, Shandong University, Jinan, China, 2023.
37. Niu, H.; Bonati, L.; Piaggi, P.M.; Parrinello, M. Ab initio phase diagram and nucleation of gallium. *Nat. Commun.* **2020**, *11*, 2654. [CrossRef] [PubMed]
38. Weir, C.E.; Piermarini, G.J.; Block, S. Piermarini, On the Crystal Structures of Cs II and Ga II. *J. Chem. Phys.* **1971**, *54*, 2768–2770. [CrossRef]
39. Boslo, P.L.; Cruien, H.; Dupont, M.; Rimsky, A. Structure cristalline de Gaδ. *Acta Cryst.* **1973**, *B29*, 367–368.
40. Boslo, P.L.; Defrarn, A. Structure cristalline du gallium β. *Acta Cryst.* **1969**, *B25*, 995.
41. Boslo, P.L.; Cruien, H.; Dupont, M.; Rimsky, A. Structure cristalline de Ga γ. *Acta Cryst.* **1972**, *B28*, 1974–1975.
42. Gong, X.G.; Chiarotti, G.L.; Parrinello, M.; Tosatti, E. α-gallium: A metallic molecular crystal. *Phys. Rev. B* **1991**, *43*, 14277–14280. [CrossRef] [PubMed]
43. Heine, V. Crystal structure of gallium metal. *J. Phys. C Solid State Phys.* **1968**, *1*, 222–231. [CrossRef]
44. Tran, R.; Xu, Z.; Radhakrishnan, B.; Winston, D.; Sun, W.; Persson, K.A.; Ong, S.P. Surface energies of elemental crystals. *Sci Data* **2016**, *3*, 160080. [CrossRef]

materials

MDPI

Article

First-Principles Study on the Electrical and Thermal Conductivities of Cu–Zn Binary Alloys

Lei Huang [1], Bo Peng [2], Qinchi Yue [2], Guojie Huang [1,*], Changhao Wang [2,*], Ruzhi Wang [2] and Ning Tian [1]

[1] High-Performance Copper Alloy Materials R&D Division, China Nonferrous Metals Innovation Institute (Tianjin) Co., Ltd., Tianjin 300393, China; huanglei.my@foxmail.com (L.H.); tianningnene@163.com (N.T.)

[2] State Key Laboratory of Materials Low-Carbon Recycling, College of Material Science and Engineering, Beijing University of Technology, Beijing 100124, China; pengbo@emails.bjut.edu.cn (B.P.); yueqinchi@emails.bjut.edu.cn (Q.Y.); wrz@bjut.edu.cn (R.W.)

* Correspondence: huangguojiepaper@126.com (G.H.); wangch33@bjut.edu.cn (C.W.); Tel.: +86-13520406080 (G.H.); +86-15801287220 (C.W.)

Abstract: Cu–Zn alloys are widely used engineering materials with well-known industrial applications. However, studies on their electrical and thermal conductivities have primarily relied on experimental measurements, while theoretical investigations remain limited. In this work, eight crystal structure models were constructed to represent three phase configurations (α single phase, $\alpha + \beta'$ dual phase, and β' single phase) of Cu–Zn alloys with Zn concentrations ranging from 0 to 50 at.%. Based on the first-principles calculations combined with the Boltzmann transport equation, the electrical and thermal conductivities of these models were computed, and the electronic structure of the α-phase configurations was further analyzed. The results show that both electrical and thermal conductivities exhibit a non-monotonic trend with increasing Zn content, initially decreasing and then increasing. This trend is in strong agreement with available experimental data. Further analysis of the electronic structure reveals that, in the α-phase region, the density of states near the Fermi level is mainly contributed by Cu d-orbitals. As Zn content increases, the effective DOS near the Fermi level decreases, leading to reduced electron transport capability. For thermal conductivity, both the Wiedemann–Franz law and the first-principles calculations were employed, yielding results consistent with experimental trends. In summary, this study systematically investigates the variation of electrical and thermal conductivities of Cu–Zn binary alloys with Zn content and explores the underlying physical mechanisms from the perspective of electronic structure. The findings provide valuable theoretical support for understanding and optimizing the transport properties of complex alloy systems.

Keywords: Cu–Zn alloy; first-principles calculations; electrical conductivity; thermal conductivity

1. Introduction

Copper–zinc (Cu–Zn) alloys are widely used in modern industry due to their high electrical and thermal conductivity, excellent corrosion resistance, and good workability. These properties make them attractive for applications in electronics, automotive manufacturing, and piping systems. In the electronics industry, Cu–Zn alloys are commonly used in connectors and terminals for their high conductivity and thermal stability. In the field of automobile manufacturing, their superior thermal conductivity and corrosion resistance make them suitable for radiator cores, connectors, and related components [1–3].

The Cu–Zn alloy system exhibits complex phase transformation behavior depending on zinc content and temperature. The maximum solid solubility of Zn in Cu is approximately 38.95 wt.%. When the Zn content is below this limit, the alloy primarily shows a face-centered cubic (FCC) α-phase. As the Zn content exceeds 40 wt.%, the alloy enters the α + β two-phase region [4]. In this range, the β-phase exists as a disordered body-centered cubic (BCC) structure above 454 °C and transforms into an ordered β′-phase upon cooling below this temperature. The β′-phase retains the BCC lattice but exhibits long-range order, such as DO3 or B2-type structures [5]. With further increase in Zn content to around 50 wt.%, the β-phase becomes dominant, forming a single-phase β structure [6].

Beyond 50 wt.% Zn, additional complex intermetallic phases such as γ, δ, and ε emerge, characterized by more intricate crystal structures, significant atomic disorder, and vacancy-induced lattice relaxations. These features fundamentally change the nature of the phase transformation behavior compared to the Cu-rich side. These phase transformations significantly influence the electrical, thermal, and mechanical properties of Cu–Zn alloys [7–11]. From an industrial perspective, the compositional range of 0–50 wt.% Zn covers the most commercially important brasses. According to international standards such as ASTM B36 [12], widely used wrought and cast brass grades (C21000–C28000) typically contain 37–41 wt.% Zn. Alloys with Zn content above 50 wt.% suffer from severe deterioration in workability and electrical conductivity, making them unsuitable for widespread industrial applications. By adjusting the phase structure and composition, the physical properties of the alloys can be tailored, which is valuable for applications in electronics, electrical engineering, and structural components.

With the development of computational materials science, the first-principles calculations based on the density functional theory (DFT) have become widely used in metallic materials research. These methods enable the investigation of materials at the electronic structure level and have provided important insights into phase stability, electronic properties, and transformation mechanisms.

For the Cu-rich side (Zn < 50 wt.%) of the Cu–Zn system, first-principles calculations have successfully been applied to analyze order–disorder transitions between the α and β phases under various compositions and temperatures. The phase stability and transformation mechanisms have been explained from an electronic structure perspective, offering a clear understanding of their stability and evolution with composition and temperature [13–18]. However, modeling the Zn-rich side (Zn > 50 wt.%) is considerably more challenging. The presence of γ, δ, and ε phases introduces highly complex, vacancy-mediated structures and significant lattice distortions, making it difficult to construct accurate atomic models and maintain consistent evaluation criteria across different phases [11]. Moreover, the underlying physical mechanisms governing phase transformations in the Zn-rich region differ substantially from those in the Cu-rich region, further complicating theoretical investigations. Similar phase-change behavior upon crossing 50 at.% composition has also been observed in other alloy systems, such as in Sn–Sb and Cu–Ni–Cr-based materials, where doping-induced structural transitions lead to significant changes in physical properties [19,20]. Considering these factors, this study focuses on Cu–Zn alloys with Zn contents up to 50 wt.%.

Experimental data have also revealed that the electrical and thermal conductivity of Cu–Zn alloys varies with Zn content [8,9]. Specifically, both properties tend to decrease initially and then increase as Zn content rises from 0 to 50 wt.%. However, most of these findings rely on experimental methods. The lack of theoretical modeling limits understanding of the underlying mechanisms. Moreover, experimental approaches often require substantial effort in sample preparation and testing and may introduce inconsistencies due to varying conditions.

In this context, atomic models of Cu–Zn alloys in the α, α + β′, and β′ phase regions were constructed. The electrical and thermal transport properties, along with the electronic structures, are investigated using DFT combined with the Boltzmann transport equation. The influence of phase structure and composition on transport properties is analyzed in detail. This work aims to uncover the mechanisms by which different phase structures affect electron transport behavior. By correlating theoretical calculations with available experimental data, the study provides a deeper understanding of the relationship between electronic structure and transport performance.

Furthermore, recent studies on the band topological characteristics of alloy materials have confirmed the nontrivial topological band structure of the Sn–Sb alloy system. Its topologically protected surface state can significantly increase the carrier mobility and generate abnormal transport behaviors [21,22]. Although the current research on Cu–Zn alloys mainly focuses on the influence of traditional phase transitions and electronic structures on transport performance, topological band engineering provides a new idea for regulating their electrical/thermal conductivity. For instance, inducing topological quantum phase transitions through composition design or generating topological surface states by utilizing interface effects may break through the performance limits of traditional alloys [23]. The electronic structure-transport property correlation framework established in this study not only lays a foundational basis for exploring potential topological effects in Cu–Zn systems but also provides theoretical guidance for alloy design and multifunctional optimization.

2. Computational Methods

2.1. Computational Models

According to the Cu–Zn binary phase diagram [24], within the composition range investigated in this study (Zn $\leq$ 50 at.%), the Cu–Zn alloy system exhibits three typical phase structures with increasing Zn content: single-phase α (FCC), dual-phase α + β′, and single-phase β′ (BCC). To investigate the effects of different phases and compositions on the electrical and thermal conductivities of Cu–Zn alloys, the crystal structure models were constructed for Zn concentrations ranging from 0 to 50 at.%.

In the α phase, Zn atoms randomly substitute Cu atoms in a disordered solid solution with a face-centered cubic (FCC) structure. To accurately represent the atomic disorder, the alloy theoretic automated toolkit (ATAT) [25,26] was used to generate special quasi-random structures (SQS). Six α-phase structures with different Zn concentrations (12.5, 18.75, 21.875, 25, 28.125, and 31.25 at.%) were constructed, as shown in Figure 1a–f. For each concentration, 20 α-phase SQS models were generated, and their ground-state energies were calculated. Unstable structures were eliminated, and the remaining structures were subjected to molecular dynamics simulations for diffusion-annealing treatment. The most stable configuration was selected as the representative structure for each composition. The β′ phase, a Cu–Zn intermetallic compound with a body-centered cubic (BCC) structure, has a fixed Cu:Zn atomic ratio of 1:1 and exhibits an ordered solid solution, as shown in Figure 1g. The α + β′ dual-phase structure was constructed by combining the α and β′ unit cells at a defined interface. Due to the significant lattice mismatch between the α phase (a = 3.677 Å) and the β′ phase (a = 2.921 Å), the vasp.6.3.0 software was used to reconstruct the lattice vectors. The final α + β′ interface model has an interfacial spacing of 1.8 Å and an interfacial angle $\gamma = 45°$, as shown in Figure 1h. All electronic structure and transport property calculations for the Cu–Zn alloys in this study were performed based on the above models.

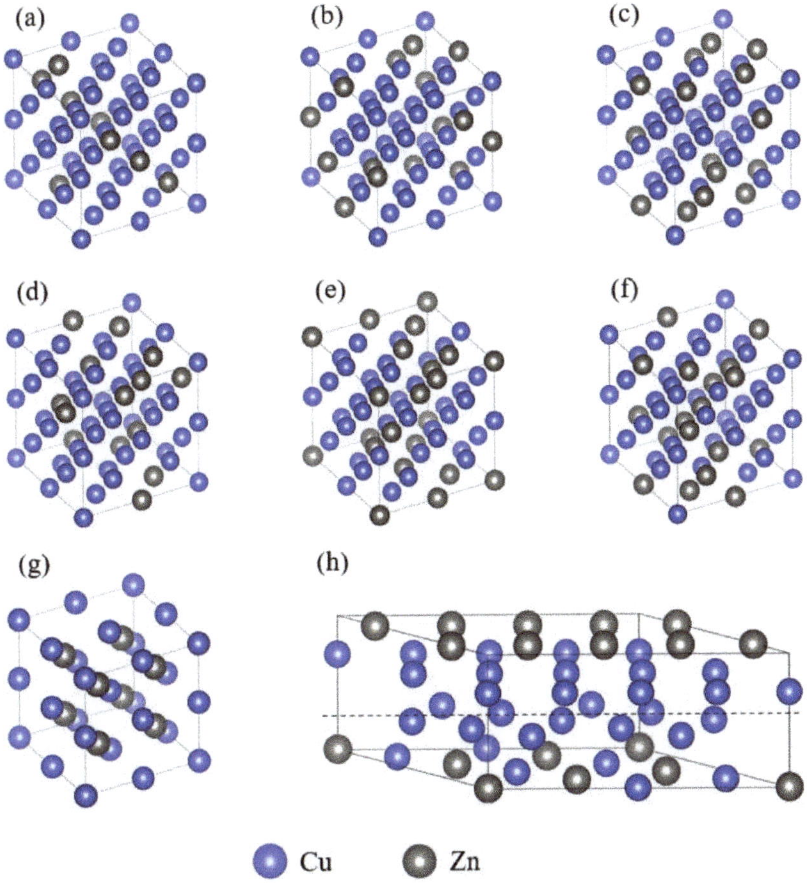

Figure 1. Crystal structure models of Cu–Zn alloys with different Zn content: (**a**) α phase with 12.5% Zn content; (**b**) α phase with 18.75% Zn content; (**c**) α phase with 21.875% Zn content; (**d**) α phase with 25% Zn content; (**e**) α phase with 28.125% Zn content; (**f**) α phase with 31.25% Zn content; (**g**) β′ phase with 50% Zn content; (**h**) α + β′ diphase configuration.

2.2. Electrical Conductivity Calculations

The electronic transport coefficients were calculated by solving the Boltzmann transport equation (BTE) under the relaxation time approximation (RTA) [27], as implemented in the BoltzTraP2 code [28]. The transport coefficients, including the electrical conductivity (σ) and the electronic thermal conductivity (κ_e), are defined as follows:

$$\sigma(T,\mu) = -\frac{\tau}{\Omega}\int_{-\infty}^{+\infty}\sigma(\varepsilon)\frac{\partial f(\varepsilon,\mu,T)}{\partial\varepsilon}d\varepsilon, \tag{1}$$

$$\kappa(T;\mu) = -\frac{\tau}{q^2 T\Omega}\int_{-\infty}^{+\infty}\sigma(\varepsilon)(\varepsilon-\mu)^2\frac{\partial f(\varepsilon,\mu,T)}{\partial\varepsilon}d\varepsilon, \tag{2}$$

where τ is the relaxation time, Ω is the unit cell volume, and f is the Fermi–Dirac distribution function, which is defined as follows:

$$f(\varepsilon) = \frac{1}{e^{\frac{\varepsilon-\mu}{kT}}+1}, \tag{3}$$

with ε being the electron energy, μ the chemical potential, k the Boltzmann constant, and T the absolute temperature. The transport distribution function (TDF) $\sigma(\varepsilon)$ is expressed as follows:

$$\sigma(\varepsilon) = \frac{1}{N}\sum_{ik} \sigma_{\alpha\beta}(i,k)\delta(\varepsilon - \varepsilon_k), \quad (4)$$

where N is the number of k-points and σ_{ik} is the conductivity tensor at each k-point [27,29], calculated as follows:

$$\sigma_{\alpha\beta}(i,k) = e^2\tau_{i,k}v_\alpha(i,k)v_\beta(i,k), \quad (5)$$

where e is the elementary charge, and v_a and v_β are the group velocities along the α and β directions, given by $(1/\hbar)$ $(\partial\epsilon/\partial k)$. BoltzTraP2 uses Fourier interpolation of the band structure obtained from VASP calculations to generate denser k-point meshes, enabling more accurate and efficient evaluation of the transport properties [30].

2.3. Thermal Conductivity Calculations

The thermal conductivity (κ) is a key parameter characterizing the heat transport capability of materials and includes contributions from both electronic (κ_e) and lattice (phonon) thermal conductivities (κ_{ph}) [31], calculated as follows:

$$\kappa = \kappa_e + \kappa_{ph}, \quad (6)$$

For metals where heat transport is predominantly electronic, κ_e can be estimated from the electrical conductivity using the Wiedemann–Franz law [32], calculated as follows:

$$\kappa_e = L\sigma T, \quad (7)$$

where L is the Lorenz number (typically approximated as 2.44×10^{-8}WΩK^{-2}), σ is the electrical conductivity, and T is the absolute temperature.

2.4. Computational Parameters

All calculations were performed using the Vienna Ab initio Simulation Package (vasp.6.3.0) [33,34]. The projector augmented wave (PAW) method [35,36] was employed, and the exchange-correlation interactions were described using the Perdew–Burke–Ernzerhof (PBE) functional. A plane-wave energy cutoff of 400 eV was used. The total energy convergence criterion was set to 1.0×10^{-6} eV/atom.

k-point sampling in the Brillouin zone was performed using the Monkhorst–Pack scheme centered at the Γ point. For structure optimization, the k-point meshes were $4 \times 4 \times 4$ for the α phase, $12 \times 12 \times 12$ for the β' phase, and $4 \times 3 \times 5$ for the $\alpha + \beta'$ dual-phase models. Geometry optimizations were carried out using the conjugate gradient algorithm [37], and the convergence criterion for atomic forces was 0.02 eV/Å.

All total energy and electronic structure calculations were based on the optimized structures. The k-point meshes used for density of states (DOS) calculations were the same as the geometry optimizations.

3. Results and Discussion

3.1. Structural Optimization Results

To verify the reliability and accuracy of our calculations, the optimized lattice parameters of pure Cu (FCC), Zn (HCP), and the Cu–Zn alloy α-phase (FCC) and β'-phase (BCC) were compared with experimental values, as shown in Table 1. The results indicate that the optimized lattice constants in this study are in good agreement with the experimental data reported in the literature [38–42]. The deviations in lattice parameters are within 1%, demonstrating the reasonableness and credibility of our structural optimization results.

Table 1. Calculated lattice parameters and experimental crystallographic data of pure copper, pure zinc, and Cu–Zn intermetallic compounds.

Phase	Crystal Structure	Lattice Parameters (Å)		Ref.
Cu	FCC	$a = 3.6077$		Exp. [40]
		$a = 3.6072$		Exp. [41]
		$a = 3.615$		This study
Cu_3Zn-α	FCC	$a = 3.6718$		Exp. [43]
		$a = 3.677$		This study
CuZn-β′	BCC	$a = 2.9575$		Exp. [44]
		$a = 2.921$		This study
Zn	HCP	$a = 2.6594$	$c = 4.9368$	Exp. [42]
		$a = 2.6190$	$c = 4.986$	This study

3.2. Electrical Conductivity: Results and Discussion

As an important industrial alloy, Cu–Zn has been extensively studied, and considerable experimental data on its properties have been accumulated. In particular, Aoyama, S., and Ito, T., conducted a series of electrical conductivity measurements on Cu-Zn alloys in the 1940s [8], which are used here as reference standards for validating our computationa results.

The nonequilibrium behavior of conduction electrons in metals can be described by the Boltzmann transport equation (BTE), which relates the response of electrons to external forces with changes in their distribution function. Due to the complexity of electron scattering processes, the relaxation time approximation (RTA) is often employed to simplify the BTE by introducing an average timescale for scattering. This approximation assumes that electrons reach a quasi-equilibrium state between scattering events, allowing their transport behavior to be described using group velocity and relaxation time.

In this work, electrical conductivity is calculated by combining the Fermi-Dirac distribution with the BTE under the RTA framework. The resulting conductivity depends on both the group velocity and the relaxation time. The group velocity characterizes the electron propagation at a given wavevector and is calculated as the gradient of the band energy with respect to the wavevector. The relaxation time reflects the dynamics of electron scattering. In this study, a constant relaxation time $\tau = 2.375 \times 10^{-14}$ s is adopted based on the average values reported for Cu–Zn alloys in the literature [43]. The group velocity is obtained from the electronic band structures computed by VASP and further processed using BoltzTraP2. The computed and experimental conductivity values are summarized in Table 2.

Table 2. Experimental [8] and calculated conductivity of Cu–Zn binary alloy.

Zn at.%	**12.81**	**17.02**	**19.81**	**24.04**	**29.41**	**35.3**	**39.39**	**43.68**	**48.02**
Conductivity (experimental) 10^7 S/m	3.56	3.08	3.04	2.83	2.46	2.39	3.28	4.73	8.25
Zn at.%	12.5	18.75	21.875	25	28.125	31.25	36	-	50
Conductivity (computed) 10^7 S/m	2.02	1.83	1.78	1.66	1.57	1.49	1.02	-	4.46

Figure 2 compares the experimental and calculated electrical conductivities of Cu–Zn alloys as a function of Zn concentration. Overall, both data sets exhibit excellent agreement in terms of trends and conductivity variations across different phase regions, validating the effectiveness of the first-principles approach. In the α-phase region (FCC random solid solution, Zn < 35 at.%), both experimental and theoretical results show a decreasing conductivity with increasing Zn content. This behavior is attributed to the increased lattice distortion caused by the random substitution of Zn atoms, which enhances electron scattering, shortens the mean free path of conduction electrons, and consequently reduces electrical conductivity.

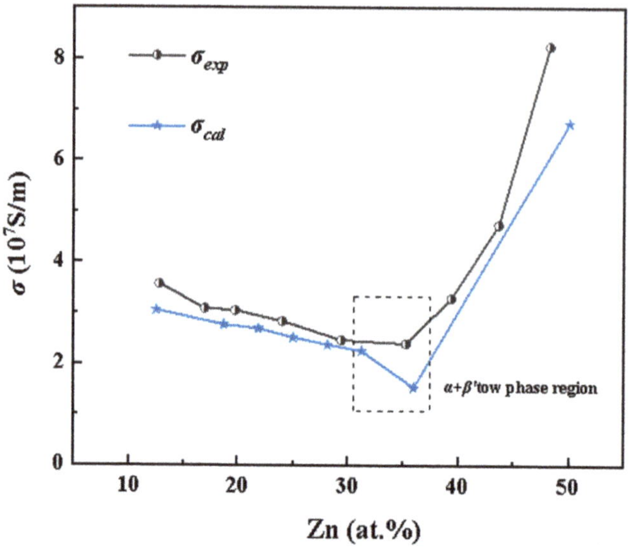

Figure 2. Curve of Cu–Zn alloy conductivity with Zn content, where the blue star curve (σ_{cal}) is the calculated results, and the black dot curve (σ_{exp}) is the experimental results [8].

In the $\alpha + \beta'$ two-phase region (Zn = 35–50 at.%), the β' phase (ordered BCC structure) begins to form and increases in proportion with Zn content. Experimentally, as the alloy transitions from the α single-phase to the $\alpha + \beta'$ two-phase region, the conductivity trend levels off, followed by a noticeable increase when Zn exceeds 37.04 at.%. However, a slight drop in the calculated conductivity at 36 at.% Zn is observed. This discrepancy is likely due to the simplifications in the two-phase structural model, which cannot fully capture the complex interfacial features present in real materials, thereby affecting the accuracy of the calculated conductivity.

At approximately 50 at.% Zn, the alloy enters the β' single-phase region, where the experimental conductivity reaches 8.25×10^7 S/m, significantly higher than that of the α-phase region. The calculated results show excellent agreement with this experimental trend. The enhanced conductivity in the β' phase is attributed to its ordered BCC structure, which reduces electron scattering and improves the continuity of electron transport by minimizing lattice imperfections.

In summary, the calculated electrical conductivities of Cu–Zn alloys using the first-principles calculations show good agreement with experimental data across different phases and compositions. This confirms the feasibility of using first-principles approaches to predict electrical conductivity variations in metal alloys.

3.3. *Effect of Zn Content on the Electronic Structure*

To gain deeper insight into the observed decrease in electrical conductivity of the α-phase Cu–Zn alloy with increasing Zn content, we analyzed the electronic structure by calculating the density of states (DOS) at various Zn concentrations. The DOS describes the number of available electronic states per unit energy interval. In a given metal phase, small compositional changes can lead to variations in the DOS near the Fermi level. In Cu–Zn alloys, the gradual incorporation of Zn alters the electronic state distribution, which in turn affects electron transport properties. This enables us to quantify the influence of Zn content on conductivity from an electronic structure perspective.

Figure 3 presents the total density of states (TDOS) and the projected density of states (PDOS) of Cu and Zn d-orbitals for α-phase Cu–Zn alloys at 12.5 at.% and 31.25 at.% Zn. The contributions from s- and p-orbitals are comparatively small; a detailed discussion of their characteristics is provided in Figure S1 of the Supplemental Materials.

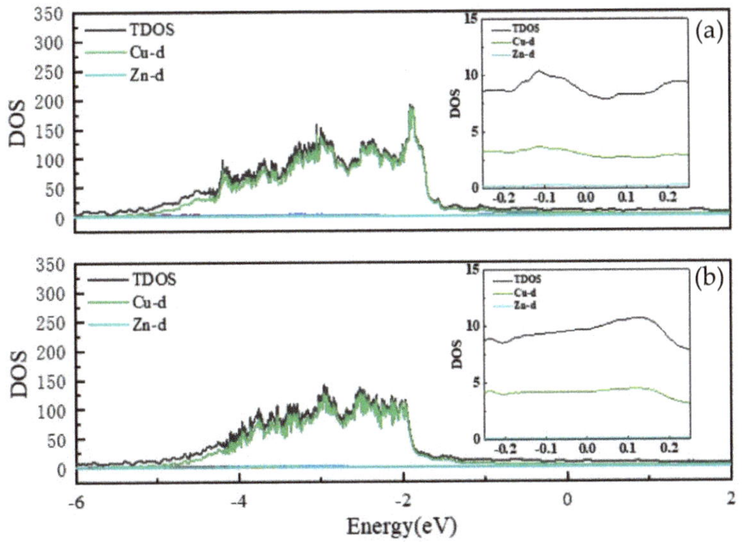

Figure 3. Total state density (TDOS) of the α phase of Cu–Zn alloy and partial wave state density (PDOS) of d-orbitals of Cu and Zn elements. (**a**) 12.5% Zn content; (**b**) 31.25% Zn content. The small graph on the right is a local magnification of the total state density near the Fermi surface.

Additionally, since the Zn d-orbitals are mainly localized between −8 eV and −7 eV, no prominent Zn d peaks appear in the energy range of interest. As observed in Figure 3, the DOS near the Fermi level is dominated by Cu d-orbitals, which serve as the primary contributors to electrical conduction.

The TDOS values near the Fermi level for six α-phase structures are listed in Table 3. These values are obtained by integrating the TDOS in the energy window from −0.25 eV to 0.25 eV. The results indicate a clear decreasing trend in the DOS at the Fermi level with increasing Zn content, which aligns with the observed decline in electrical conductivity. This trend can be attributed to the reduced number of available electronic states near the Fermi level upon Zn addition, thereby hindering electron transport and decreasing conductivity. Although Zn s- and p-orbitals provide some contribution near the Fermi level, their DOS is significantly lower than that of Cu d-orbitals, resulting in a negligible impact on the overall conductivity.

Table 3. TDOS integrals of six α phases of Cu–Zn alloy between −0.25 eV and 0.25 eV near the Fermi surface.

Zn Doping Ratio	12.5 at.%	18.75 at.%	21.875 at.%	25 at.%	28.125 at.%	31.25 at.%
Integrated TDOS Value	1.96	1.93	1.89	1.85	1.83	1.78

3.4. Thermal Conductivity: Results and Discussion

Aoyama, S., and Ito, T., experimentally measured the thermal conductivity of Cu–Zn alloys in 1940 [8], and their data are used in this study as a reference standard for evaluating the calculated results. We investigated the variation in thermal conductivity of Cu–Zn alloys with Zn content and phase transitions between the α and β' phases. Two different methods were employed to calculate the thermal conductivity at 78 K: the Wiedemann–Franz law and a first-principles approach combining density functional theory (DFT) with the Boltzmann transport theory. The corresponding results are summarized in Table 4. In the table, σ represents the experimentally measured electrical conductivity at 78 K, κ_{exp} denotes the experimental thermal conductivity, κ_{WF} is the thermal conductivity derived using the Wiedemann–Franz law, and κ_{DFT} refers to the result obtained from DFT-based calculations.

Table 4. Thermal conductivity of Cu–Zn binary alloy at 78 K.

Results of Wiedemann–Franz Law Calculation Method								
Zn at.%	17.01	19.92	24.04	29.41	35.30	39.39	43.68	48.20
κ_{exp} (W/m/K)	71	69	62	54	52	64	90	156
κ_{WF} (W/m/K)	38.82	38.31	35.67	31.01	30.12	41.34	59.62	103.98
Results of DFT calculation method								
Zn at.%	12.5	18.75	21.875	25	28.125	31.25	36	50
κ_{DFT} (W/m/K)	37.74	34.97	33.92	31.26	30.83	29.73	19.55	80.99

Figure 4 illustrates the thermal conductivity of Cu–Zn alloys at 78 K, based on the data in Table 4. The traditional method for calculating lattice (phonon) thermal conductivity involves solving the phonon Boltzmann transport equation using third-order interatomic force constants [44,45]. However, determining these third-order force constants becomes increasingly complex and computationally demanding as the number of atoms in the primitive cell increases. Moreover, the lattice contribution to the total thermal conductivity in Cu–Zn alloys has been reported to be relatively minor, typically accounting for only 4–7% of the total thermal conductivity [31]. Therefore, it is reasonable to neglect the phonon contribution in this study.

The experimental thermal conductivity values (κ_{exp}) are slightly higher than those predicted by the Wiedemann–Franz law (κ_{WF}), yet they are very close in magnitude. This suggests that although the Wiedemann–Franz approach omits the phonon contribution, it still provides a reasonably accurate estimate of the total thermal conductivity in Cu–Zn alloys. This further confirms that the electronic component dominates the thermal transport in these alloys.

Comparison between the experimental data (κ_{exp}) and the DFT-calculated results (κ_{DFT}) shows that κ_{exp} is slightly higher than κ_{DFT}, but the values are close and follow a consistent trend. This agreement among κ_{exp}, κ_{WF}, and κ_{DFT} supports the conclusion that electron thermal conductivity is the dominant contributor to the total thermal conductivity in Cu–Zn alloys and also validates the accuracy and reliability of the thermal conductivity calculations presented in this study.

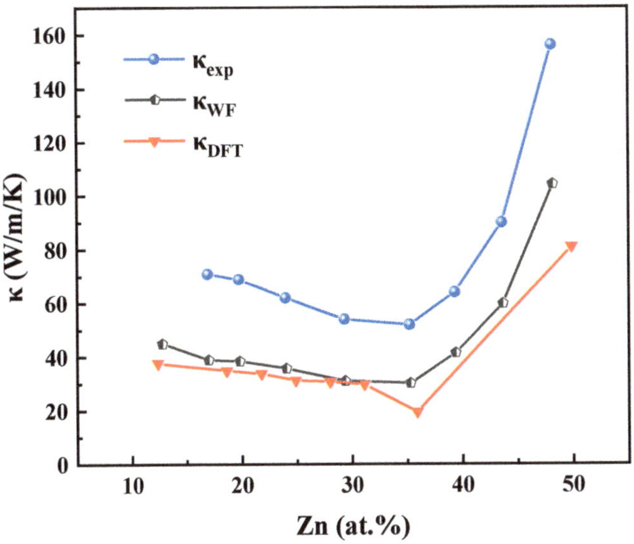

Figure 4. The thermal conductivity curves of the Cu–Zn binary alloy with different Zn contents at 78 K of the experimental data (κ_{exp}), Wiedemann–Franz Law calculation results (κ_{WF}), and DFT calculation results (κ_{DFT}).

By comparing the electrical conductivity and thermal conductivity curves of Cu–Zn alloys in Figures 2 and 4, a clear positive correlation between the two properties with respect to Zn atomic concentration can be observed. When the Zn content is below 35 at.%, both electrical and thermal conductivities decrease with increasing Zn concentration, reaching their minimum at approximately 35 at.% Zn. Beyond this point, both properties exhibit a rapid increase as Zn content continues to rise. The two sets of curves demonstrate a highly synchronized trend. This behavior reflects the dominant role of electronic transport characteristics in determining both the electrical and thermal conductivities of Cu–Zn alloys. In summary, the accuracy of electrical conductivity calculations directly influences the reliability of thermal conductivity predictions. Therefore, improving the precision of electrical conductivity calculations is essential for enhancing the overall accuracy of thermal conductivity modeling in computational studies.

4. Conclusions

In this study, eight crystal structure models of Cu–Zn binary alloys were constructed, covering three types of phase configurations within the Zn concentration range of 0–50 at.%: α single phase (FCC), $\alpha + \beta'$ dual phase, and β' single phase (BCC). The electrical and thermal conductivities of these models were calculated using the first-principles calculations in combination with the Boltzmann transport equation, and the electronic structure of the α-phase configurations was further investigated. The results show that both electrical and thermal conductivities of Cu–Zn alloys exhibit a decreasing–increasing trend with increasing Zn content from 0 to 50 at.%, displaying a clear positive correlation between the two properties. This trend is in good agreement with experimental observations. Electronic structure analysis further reveals that, in the α-phase region, the density of states (DOS) near the Fermi level is mainly contributed by the Cu d-orbitals. As Zn content increases, the effective DOS near the Fermi level decreases, leading to a reduction in electron transport capability and, consequently, a decline in electrical conductivity within the α-phase region. Regarding thermal conductivity calculations, both the Wiedemann–Franz law and first-principles methods were employed, and the calculated results exhibit trends consistent with experimental data. In summary, this work systematically investigates the variation of electrical and thermal transport properties of Cu–Zn binary alloys with Zn content and elucidates the underlying physical mechanisms from the perspective of electronic structure. The findings provide theoretical insights to support transport property studies and performance optimization in complex alloy systems.

Supplementary Materials: The following supporting information can be downloaded at: https://www.mdpi.com/article/10.3390/ma18102310/s1, Figure S1: Calculated PDOS of Zn-s and Zn-p in the structures of (a) β' single phase and (b) α single phase(Zn 12.5% content).

Author Contributions: Conceptualization, L.H. and G.H.; Methodology, B.P., Q.Y., G.H. and C.W.; Software, B.P. and Q.Y.; Validation, C.W. and N.T.; Formal analysis, B.P. and N.T.; Investigation, R.W.; Resources, L.H.; Data curation, R.W.; Writing—original draft, L.H.; Writing—review & editing, C.W.; Visualization, R.W. and N.T.; Supervision, G.H. and C.W.; Project administration, G.H.; Funding acquisition, L.H. All authors have read and agreed to the published version of the manuscript.

Funding: This research was funded by the Key Research and Development Plan of Ningbo's "Science and Technology Innovation Yongjiang 2035", grant number: 2024Z088 and the Science and Technology Project of China Nonferrous Metal Mining Group (grant number: 2023KJZX007).

Institutional Review Board Statement: Not applicable.

Informed Consent Statement: Not applicable.

Data Availability Statement: The original contributions presented in this study are included in the article/Supplementary Materials. Further inquiries can be directed to the corresponding authors.

Conflicts of Interest: Author Lei Huang, Guojie Huang and Ning Tian was employed by the company China Nonferrous Metals Innovation Institute (Tianjin) Co., Ltd. The remaining authors declare that the research was conducted in the absence of any commercial or financial relationships that could be construed as a potential conflict of interest.

References

1. Berndorf, S.; Markelov, A.; Guk, S.; Mandel, M.; Krüger, L.; Prahl, U. Development of a Dezincification-Free Alloy System for the Manufacturing of Brass Instruments. *Metals* **2024**, *14*, 800. [CrossRef]
2. Brans, K.; Kind, S.; Meurer, M.; Bergs, T. Influence of the Material Production Route on the Material Properties and the Machinability of the Lead-Free Copper-Zinc-Alloy CuZn40 (CW509L). *Metals* **2024**, *14*, 747. [CrossRef]
3. Stavroulakis, P.; Toulfatzis, A.I.; Pantazopoulos, G.A.; Paipetis, A.S. Machinable leaded and eco-friendly brass alloys for high performance manufacturing processes: A critical review. *Metals* **2022**, *12*, 246. [CrossRef]
4. Suárez, L.; Rodriguez-Calvillo, P.; Cabrera, J.; Martínez-Romay, A.; Majuelos-Mallorquín, D.; Coma, A. Hot working analysis of a CuZn40Pb2 brass on the monophasic (β) and intercritical (α+ β) regions. *Mater. Sci. Eng. A* **2015**, *627*, 42–50. [CrossRef]
5. Garg, R.; Ranganathan, S.; Suwas, S. Effect of mode of rolling on development of texture and microstructure in two-phase (α+ β) brass. *Mater. Sci. Eng. A* **2010**, *527*, 4582–4592. [CrossRef]
6. Yu, Y.; Chen, W.; Xie, L.; Song, Y. Effect of heat treatment on microstructure evolution, phase transformation and mechanical properties of dual phase Cu-Zn alloy. *J. Alloys Compd.* **2022**, *904*, 163960. [CrossRef]
7. Choi, G.; Kim, H.S.; Lee, K.; Park, S.H.; Cha, J.; Chung, I.; Lee, W.B. Study on thermal conductivity and electrical resistivity of Al-Cu alloys obtained by Boltzmann transport equation and first-principles simulation: Semi-empirical approach. *J. Alloys Compd.* **2017**, *727*, 1237–1242. [CrossRef]
8. Ho, C.; Ackerman, M.; Wu, K.; Oh, S.; Havill, T. Thermal conductivity of ten selected binary alloy systems. *J. Phys. Chem. Ref. Data* **1978**, *7*, 959–1178. [CrossRef]
9. Ho, C.Y.; Ackerman, M.; Wu, K.; Havill, T.; Bogaard, R.; Matula, R.; Oh, S.; James, H. Electrical resistivity of ten selected binary alloy systems. *J. Phys. Chem. Ref. Data* **1983**, *12*, 183–322. [CrossRef]
10. Jang, H.W.; Hong, J.-W. Influence of zinc content on the mechanical behaviors of Cu-Zn alloys by molecular dynamics. *Materials* **2020**, *13*, 2062. [CrossRef]
11. Kong, G.-X.; Ma, X.-J.; Liu, Q.-J.; Li, Y.; Liu, Z.-T.; Matter, C. Structural stability, elastic and thermodynamic properties of Au–Cu alloys from first-principles calculations. *Phys. B Condens. Matter* **2018**, *533*, 58–62. [CrossRef]

12. *ASTM B36/B36M-23 (2023)*; Standard Specification for Brass Plate, Sheet, Strip, and Rolled Bar. ASTM International: West Conshohocken, PA, USA, 2023. Available online: https://cdn.standards.iteh.ai/samples/116325/cb60469514a0475baecfb788ef9a3d14/ASTM-B36-B36M-23.pdf (accessed on 9 May 2025).
13. Dhaka, R.; Banik, S.; Shukla, A.; Vyas, V.; Chakrabarti, A.; Barman, S.; Ahuja, B.; Sharma, B. Electronic structure of α-and β-brass. *Phys. Rev. B Condens. Matter Mater. Phys.* **2008**, *78*, 073107. [CrossRef]
14. Hong, H.; Wang, Q.; Dong, C.; Liaw, P.K. Understanding the Cu-Zn brass alloys using a short-range-order cluster model: Significance of specific compositions of industrial alloys. *Sci. Rep.* **2014**, *4*, 7065. [CrossRef] [PubMed]
15. Liu, Q.; Cheng, L. Structural evolution and electronic properties of Cu-Zn alloy clusters. *J. Alloys Compd.* **2019**, *771*, 762–768. [CrossRef]
16. Müller, S.; Zunger, A. Structure of ordered and disordered α-brass. *Phys. Rev. B* **2001**, *63*, 094204. [CrossRef]
17. Temmerman, W.; Durham, P.; Szotek, Z.; Sob, M.; Larsson, C. Electronic effects in the order-disorder transition in β brass. *J. Phys. F Met. Phys.* **1988**, *18*, 2387. [CrossRef]
18. Turchi, P.; Sluiter, M.; Pinski, F.; Johnson, D.D.; Nicholson, D.; Stocks, G.; Staunton, J. First-principles study of phase stability in Cu-Zn substitutional alloys. *Phys. Rev. Lett.* **1991**, *67*, 1779. [CrossRef]
19. Upreti, D.; Basnet, R.; Sharma, M.M.; Chhetri, S.K.; Acharya, G.; Nabi, M.R.U.; Sakon, J.; Benamara, M.; Mortazavi, M.; Hu, J. Medium-Entropy Engineering of Magnetism in Layered Antiferromagnet CuxNi2(1−x)CrxP2S6. *Adv. Funct. Mater.* **2024**, *35*, 24018722. [CrossRef]
20. Sharma, M.M.; Kumar, K.; Sang, L.; Wang, X.L.; Awana, V.P.S. Type-II Superconductivity Below 4 K in $Sn_{0.4}Sb_{0.6}$. *J. Alloys Compd.* **2020**, *844*, 156140. [CrossRef]
21. Li, G.; Fu, C.; Shi, W.; Jiao, L.; Wu, J.; Yang, Q.; Saha, R.; Kamminga, M.E.; Srivastava, A.K.; Liu, E.; et al. Dirac nodal arc semimetal PtSn4: An ideal platform for understanding surface properties and catalysis for hydrogen evolution. *Angew. Chem.* **2019**, *131*, 13241–13246. [CrossRef]
22. Li, H.; Zhou, D.; He, Q.; Si, N.; Xin, B.; Bu, S.; Ji, Q.; Li, H.; Fuchs, H.; Niu, T. Experimental realization and phase engineering of a two-dimensional SnSb binary honeycomb lattice. *ACS Nano* **2021**, *15*, 16335–16343. [CrossRef]
23. Li, G.; Xu, Q.; Shi, W.; Fu, C.; Jiao, L.; Kamminga, M.E.; Yu, M.; Tüysüz, H.; Kumar, N.; Süß, V.; et al. Surface states in bulk single crystal of topological semimetal Co3Sn2S2 toward water oxidation. *Sci. Adv.* **2019**, *5*, eaaw9867. [CrossRef] [PubMed]
24. Eastman, C.M.; Zhang, Q.; Zhao, J.-C. Diffusion coefficients and phase equilibria of the Cu-Zn binary system studied using diffusion couples. *J. Phase Equilibria Diffus.* **2020**, *41*, 642–653. [CrossRef]
25. Van De Walle, A. Multicomponent multisublattice alloys, nonconfigurational entropy and other additions to the Alloy Theoretic Automated Toolkit. *Calphad* **2009**, *33*, 266–278. [CrossRef]
26. van de Walle, A.; Tiwary, P.; de Jong, M.; Olmsted, D.L.; Asta, M.; Dick, A.; Shin, D.; Wang, Y.; Chen, L.-Q.; Liu, Z.-K. Efficient stochastic generation of special quasirandom structures. *Calphad* **2013**, *42*, 13–18. [CrossRef]
27. Mahan, G.; Sofo, J. The best thermoelectric. *Proc. Natl. Acad. Sci. USA* **1996**, *93*, 7436–7439. [CrossRef]
28. Madsen, G.K.; Singh, D.J. BoltzTraP. A code for calculating band-structure dependent quantities. *Comput. Phys. Commun.* **2006**, *175*, 67–71. [CrossRef]
29. Van der Laag, N.J.; Fang, C.M.; De With, G.; De Wijs, G.A.; Brongersma, H.H. Geometry of {001} Surfaces of Spinel (MgAl2O4): First-Principles Simulations and Experimental Measurements. *J. Am. Ceram. Soc.* **2005**, *88*, 1544–1548. [CrossRef]
30. Madsen, G.K.; Carrete, J.; Verstraete, M.J. BoltzTraP2, a program for interpolating band structures and calculating semi-classical transport coefficients. *Comput. Phys. Commun.* **2018**, *231*, 140–145. [CrossRef]
31. Cui, Y.; Li, S.; Ying, T.; Bao, H.; Zeng, X. Research on the thermal conductivity of metals based on first principles. *Acta Met. Sin* **2020**, *57*, 375–384.
32. Lavasani, A.; Bulmash, D.; Das Sarma, S. Wiedemann-Franz law and Fermi liquids. *Phys. Rev. B* **2019**, *99*, 085104. [CrossRef]
33. Kresse, G.; Furthmüller, J. Efficiency of ab-initio total energy calculations for metals and semiconductors using a plane-wave basis set. *Comput. Mater. Sci.* **1996**, *6*, 15–50. [CrossRef]
34. Kresse, G.; Hafner, J. Ab initio molecular-dynamics simulation of the liquid-metal–amorphous-semiconductor transition in germanium. *Phys. Rev. B* **1994**, *49*, 14251. [CrossRef]
35. Blöchl, P.E. Projector augmented-wave method. *Phys. Rev. B* **1994**, *50*, 17953. [CrossRef]
36. Kresse, G.; Joubert, D. Projector augmented-wave method. *Phys. Rev. B* **1999**, *59*, 1758. [CrossRef]
37. Shewchuk, J.R. *Classroom Figures for the Conjugate Gradient Method Without the Agonizing Pain*; School of Computer Science Carnegie Mellon University: Pittsburgh, PA, USA, 1994.
38. Beck, L.H.; Smith, C.S. Copper-zinc constitution diagram, redetermined in the vicinity of the beta phase by means of quantitative metallography. *JOM* **1952**, *4*, 1079–1083. [CrossRef]
39. Jette, E.R.; Foote, F. Precision determination of lattice constants. *J. Chem. Phys.* **1935**, *3*, 605–616. [CrossRef]
40. Keast, V.; Ewald, J.; De Silva, K.; Cortie, M.; Monnier, B.; Cuskelly, D.; Kisi, E. Optical properties and electronic structure of the Cu–Zn brasses. *J. Alloys Compd.* **2015**, *647*, 129–135. [CrossRef]

41. Lu, S.; Chang, Y. The accurate evaluation of lattice spacings from back-reflection powder photographs. *Proc. Phys. Soc.* **1941**, *53*, 517. [CrossRef]
42. Owen, E.A.; Yates, E.L. Precision measurements of crystal parameters. *Lond. Edinb. Dublin Philos. Mag. J. Sci.* **1933**, *15*, 472–488. [CrossRef]
43. Deinzer, G.; Birner, G.; Strauch, D. Ab initio calculation of the linewidth of various phonon modes in germanium and silicon. *Phys. Rev. B* **2003**, *67*, 144304. [CrossRef]
44. Gall, D. Electron mean free path in elemental metals. *J. Appl. Phys.* **2016**, *119*, 085101. [CrossRef]
45. Wang, Z.; Wang, S.; Obukhov, S.; Vast, N.; Sjakste, J.; Tyuterev, V.; Mingo, N. Thermoelectric transport properties of silicon: Toward an ab initio approach. *Phys. Rev. B Condens. Matter Mater. Phys.* **2011**, *83*, 205208. [CrossRef]

MDPI AG
Grosspeteranlage 5
4052 Basel
Switzerland
Tel.: +41 61 683 77 34

Materials Editorial Office
E-mail: materials@mdpi.com
www.mdpi.com/journal/materials

www.ingramcontent.com/pod-product-compliance
Lightning Source LLC
Chambersburg PA
CBHW061556210926
51850CB00009B/2121
* 9 7 8 3 7 2 5 8 5 7 4 1 8 *